The Genetics of
Tribolium
and
Related Species

ADVANCES IN GENETICS

Edited by

E. W. CASPARI

DEPARTMENT OF BIOLOGY
THE UNIVERSITY OF ROCHESTER
ROCHESTER, NEW YORK

J. M. THODAY

DEPARTMENT OF GENETICS
UNIVERSITY OF CAMBRIDGE
CAMBRIDGE, ENGLAND

SUPPLEMENT 1

The Genetics of *Tribolium* and Related Species

Alexander Sokoloff

The Genetics of
Tribolium
and
Related Species

Alexander Sokoloff

Department of Genetics
University of California
Berkeley, California

and

California State College at San Bernardino
San Bernardino, California

1966

ACADEMIC PRESS   New York and London

ACADEMIC PRESS INC.
111 Fifth Avenue, New York, New York 10003

United Kingdom Edition published by
ACADEMIC PRESS INC. (LONDON) LTD.
Berkeley Square House, London W.1

LIBRARY OF CONGRESS CATALOG CARD NUMBER: 65-26408

Foreword

The initial volume of *Advances in Genetics* carries a preface by the first editor, Milislav Demerec, which indicates that the purpose of the publication is to provide "critical summaries of outstanding genetic problems, written by competent geneticists. . . . The aim is to have the articles written in such form that they will be useful as reference material for geneticists and also as a source of information to nongeneticists." This monograph achieves these purposes. It started out as an article for the serial publication but when it became apparent that the volume of material covered was so vast the editors decided to issue it as a separate supplement.

Tribolium genetics is rapidly coming of age. Only 7 years ago the number of mutants known in the two most extensively studied species, *Tribolium castaneum* and *T. confusum,* was confined to seven. In the present survey over one hundred and fifty, not counting over half a hundred others in various species of Coleoptera, are described. The doubling rate of known mutations in *Tribolium* thus appears to be of the order of 18 months.

This astonishing explosion of information on the genetics of an animal with a relatively long intergeneration span (at least as compared to the currently popular objects of genetic study or even to *Drosophila*) is largely due to the work of the author of this volume and his assistants. It is therefore entirely appropriate that a comprehensive description of the various genetic effects discovered be largely a firsthand one.

Research by other geneticists working with *Tribolium* is now rapidly expanding as witnessed by the rate of growth of the annual *Tribolium Information Bulletin* (edited by Alexander Sokoloff). Not only the workers in this field but many others should find this monograph useful as a catalogue of genes and also as a compendium of much other information on cytology, linkage relations, population genetics, gene homology in different species, and certain aspects of morphogenesis. A great variety of other valuable information is included. A reference book of this sort may be expected to stimulate even further activities in the biology of Coleoptera. The author and the editors are to be congratu-

lated on this departure from, or rather addition to, the previously established format of this serial publication. Perhaps similar undertakings
with respect to other organisms may now follow.

April, 1966   I. MICHAEL LERNER
                                                   University of California
                                                   Berkeley, California

Preface

Biologists tend to be rather conservative in the use of laboratory animals. Once they have been trained in the use of one particular organism, they seldom abandon it in favor of another one. The change is considered too costly in time and effort, and the new organism may offer enough disadvantages to make the effort worthless. Having initially worked with *Drosophila*, I learned to appreciate the numerous advantages and learned to cope with the minor disadvantages these flies offer as tools in a variety of research problems in the laboratory and in the field.

The fact that I turned to another organism in my research I view as an inevitable but happy accident. As a graduate student, although Thomas Park, my major professor, and his students used flour beetles in their research, I used three species in the *pseudoobscura* subgroup of *Drosophila*. When I graduated (about 12 years ago) Dr. Park turned over to me two undescribed mutants, black in *Tribolium castaneum* and pearl in *Latheticus oryzae*. At that time the total information on the formal genetics of *Tribolium* consisted of descriptions of five mutants in the two species *T. castaneum* and *T. confusum*, and the assumption was often made that these species were rather low in mutation rate. It was not until 1958 that these beetles began to be examined fairly carefully, although not too intensively for mutants, and it is in the last 5 years or so that a number of papers answering some fundamental questions have been published. Certain problems have been solved by using genetic strains which differ not by virtue of having genetic markers but by being either highly heterogenous or highly inbred. For example, flour beetles have been used for many years by population ecologists. The growth of *T. castaneum* and *T. confusum* in single species populations given the same medium (whole wheat flour) and varying conditions of temperature and relative humidity is well known.

Thomas Park's work is often cited to illustrate that the outcome of intraspecies competition under given environmental conditions can be indeterminate. The work of Lerner and Ho was fundamental in showing that competitive ability has a strong genetic component and where the genotype is known the outcome of competition becomes determinate. Park and his collaborators have recently reported similar results.

N. E. Morton, J. F. Crow, and H. J. Muller (*PNAS* 42, 1956) proposed a method for differentiating between mutational and balanced (segregational) loads. Using wild-type, synthetic, inbred, and reconstituted strains of *T. castaneum* and *T. confusum*, H. Levene, I. M. Lerner, A. Sokoloff, F. K. Ho, and I. R. Franklin (*PNAS* 53, 1965) computed the expressed load (*A*) and the concealed load (*B*) as well as the *B/A* ratios for these strains and found that they are of the same order of magnitude as in various species of Drosophila. There were, however, some differences in these parameters, both between the two species studied and within species between inbred and noninbred populations. Furthermore, different noninbred populations of *T. confusum* show heterogeneity in *T. confusum* at the level of inbreeding (half-sib and full-sib mating) studied. The relatively low magnitudes *B/A* ratios and the existence of unexplainable differences between the various populations has led to the conclusion that the method suggested by Morton, Crow, and Muller for discriminating mutational and balanced loads is not a particularly useful one.

Another problem which can be attacked without the use of mutant markers is the very nature of survivorship of one species or the other when *T. castaneum* and *T. confusum* are confined in vials containing measured (but renewable) food. Numerous experiments have shown that when wild-type strains are introduced the first species is the survivor and the second, *T. confusum*, becomes extinct. Recent (unpublished) work utilizing different media casts doubt on the assertion that what is being observed is competition. The interaction between the two species is rather a double predator-prey relationship, *T. castaneum* becoming the predator and *T. confusum* the prey and vice versa.

For other studies mutant material is needed, and it has suddenly become abundant. In fact, about 90% of the more than 200 mutants described here were found in the last 5 years, and in our laboratory, in the period between February and December 1965, about two dozen additional mutants were discovered.

I predict that some of the more viable and readily identifiable mutations will be useful as markers in ecological or ecogenetic problems where such material may be required. The remaining mutations, although not useful for these purposes because of incomplete penetrance, variable expression, poor viability, etc., should not be ignored since they provide clues to certain questions which in the past have been given only tentative and speculative answers. For example, the role of the odoriferous glands was long a subject of speculation. Because flour beetles belong to the Tenebrionidae, and the same family includes genera such as *Eleodes* which *do* use these glands for defensive purposes, it was assumed that these glands played a defensive role. L. M. Roth, however, pointed out that in their present habitat the only serious predators of *Tribolium* are mites, and the quinones secreted by the odoriferious glands did not prevent mites from attacking the beetle. Later, J. H. Van Wyk, A. C. Hodson, and C. M. Christensen suggested that the quinones secreted by the beetles play a fungistatic and bacteriostatic role. This was confirmed recently by the discovery of the melanotic stink glands (*msg*) mutation. Dense, wild-type cultures discolor the flour as it is used up, but the medium remains particulate because of the presence of quinones. Dense cultures of *msg*, on the other hand, unable to secrete quinones, cannot prevent the development of mold in the flour which they infest (Engelhardt, Rapoport, and Sokoloff, 1965). The problem of dispersion may be intimately related to the secretions of the glands in normal strains.

From the standpoint of evolution, one fundamental question is how organisms manage to survive when a piece or a whole chromosome is lost. Several species of *Tribolium*, investigated cytologically by S. G. Smith (1952), appear to have evolved in a stepwise fashion: from the primitive condition found in *T. castaneum* to the more advanced condition in *T. destructor* there has first been a translocation of a pair of autosomes to the X and y (the condition found in *T. confusum*), and later the elimination of much of this autosomal element from the neo-X and neo-Y (the condition found in *T. destructor*). The mutant material already available in *Tribolium* suggests the possible homology of the mutants prothoraxless (*ptl*) in *T. castaneum* and prothoraxless-like in *T. confusum*, and the autosome carrying *ptl* may be the one which was translocated to the X and y in the evolution of *T. confusum*. To enable us to explore this question further, more markers are needed, not only for these two species but for *T. destructor*.

In general, much of the mutant material in *Tribolium* has provided a genetic explanation for many of the numerous abnormalities found in beetles of diverse families and reported repeatedly as developmental accidents or teratologies. Certain homeotic mutants such as antennapedia, labiopedia, and maxillopedia (just found), and alate prothorax provide further evidence for the evolution of the various appendages in insects.

Attempts have been made to solve the problem of homology between the parts of the genital organs in the two sexes by examining them morphologically, histologically, or during metamorphosis in the pupa. The results appear to be unsatisfactory. It is believed that some of the material presented in this monograph, as well as more recently discovered mutants such as Tetraurogomphi, will provide a more satisfactory solution to this problem.

Since some linkage data are already available, *Tribolium* mutants can be used to locate insecticide resistance factors in these beetles. I do not believe it necessary to stress the importance of these mutants in teaching the principles of genetics.

In accepting with alacrity the invitation to review the status of the formal genetics of *Tribolium* and related forms I am fully aware that the "barest surface has been scratched" regarding this subject in the Coleoptera. At the same time, since this information has been published primarily in the *Tribolium Information Bulletin,* the time has come to draw the attention of geneticists, ecologists, morphologists, and students of evolution to the interesting material available in beetles, and their suitability to various phases of research. Perhaps a few will be persuaded to switch organisms in their work.

I wish to thank the editors of *Advances in Genetics,* particularly Dr. Ernst W. Caspari, for all they have done to make this publication possible; Mrs. Merlene Hohmann for typing the manuscript; and my wife Barbara who spent many lonely evenings while I devoted my time to writing.

ALEXANDER SOKOLOFF

Contents

I. Introduction

Flour beetles of the genus *Tribolium* and related genera of the Tenebrionidae constitute important primary and secondary pests in all kinds of cereal products. For this reason, they have attracted the attention of investigators with a broad spectrum of interests. The literature dealing with embryology and morphology, various physiological and biochemical aspects, effects of irradiation and insecticides, parasite and host interactions, nutrition, etc., is too vast to review here. It should, however, be pointed out that since Chapman (1924) began to use, and stressed the usefulness of, *Tribolium* in population studies, *Tribolium castaneum* (to be referred to as CS) and *T. confusum* (CF) have received considerable attention from ecologists. As a result, these are probably among the best ecologically documented species of insects in existence. Particularly extensive have been the experimental investigations by Park and collaborators on population physiology and *intra-* and interspecies competition between CS and CF (see Park *et al.*, 1964; Lerner and Ho, 1961, for other references); by Stanley on various aspects on the growth of populations and cannibalism in CF (see, for example, Stanley, 1964a,b); and by Howe (1956, 1960, 1962a,b), who has investigated various aspects of the ecology of a wide range of beetles infesting stored products, including *Tribolium castaneum*, *T. confusum*, and *T. madens*.

While ecological information on *Tribolium* has grown rapidly in the last 40 years, genetic information (as is generally true for the Coleoptera and largely true for other orders of insects except some species in the Diptera, Hymenoptera, and Lepidoptera) was limited until 1958 to the description and mode of inheritance of a total of five mutations, three in CS (Park, 1937; Miller, 1944; Park and Frank, 1951), and two in CF (Park *et al.*, 1945; Graham, 1957).

At first sight this scarcity of information might be interpreted as being the result of a lower mutation rate in beetles than in other organisms. This, as can be seen from the numbers of mutants in CS, CF, and related forms included in this review, is far from the truth. In the last 6 years, without particularly diligent searching, over one hundred mutants have been found in CS and over fifty in CF. Most of these have appeared spontaneously. A few have been found in descendants of beetles exposed to X-rays, but in view of the apparently high spontaneous mutation rate in material which has not been deliberately inbred the fact that these mutants were found following irradiation may have been coincidental.

Technically, CS and CF have many advantages to recommend their use in genetic research:

1. They can be reared quite readily in a particular medium consisting of whole wheat flour supplemented with brewer's yeast in a weight ratio of 19:1. Other, less suitable, media are readily available in commercial outlets.

2. They can be reared at a wide range of temperatures and relative humidities. CF, for example, has an optimal temperature for development of 32.5°C under high relative humidity (70%) conditions, the beetles developing in about 25 days, but they can withstand a temperature of 37.5°C at the upper limit, and 17.5°–20°C at the lower limit if relative humidity is high (Howe, 1960). CS developed fastest at 35°C and high humidity, requiring 24 days to develop from egg to adult. This species fails to develop at 24°C and 30% relative humidity and at 40°C and 90% and 30% relative humidity (Howe, 1956).

3. Since the medium in which flour beetles live consists of fine particles, all the various stages of the life cycle can be isolated from the flour by using sieves of different mesh.

4. Sex is readily determinable in the pupa by the larger genital lobes in the female and the minute ones in the male. The adult can be identified readily by the genitalia, or by the basal pit located on the proximal medial surface of the femur, which is present on the first pair of legs in CS or on all pairs in CF males, those in the first pair of legs being the largest, but still smaller and not as easily seen as those in CS (Hinton, 1942).

5. The imagoes are long-lived. Good (1936) recorded a CS male which lived for 2.5 years and a CF male which lived for 3.5 years in isolation, but the average life expectancy for CS is about 6 months, and somewhat longer for CF .

6. Females become sexually mature 1 or 2 days after eclosion, reach their maximum egg laying capacity when 5–10 days old, and with the stimulus of a male partner produce eggs at the rate of 10–20 per day; CS maintains this rate for 4–5 months, while CF maintains this rate for a longer period (see, for example, Park et al., 1958). Both species produce far greater numbers of progeny if the parents are transferred to fresh medium at short intervals of a few days (Sokoloff, 1961a; Sokoloff et al., 1965).

7. Larvae and pupae are easily extracted from the medium and the pupae are not enclosed in a chitinous envelope, so that the investigator is enabled to detect abnormalities in these stages.

8. Stocks require little care. If kept at room temperature (25°C) a culture will maintain itself for 4–6 months before it becomes necessary to replace the medium. However, certain precautions are necessary, because without proper ventilation metabolic products accumulate in the flour in dense cultures and the medium may become gummy and trap the beetles.

There are a few disadvantages in working with these beetles:

1. The motile stages (larvae and adults) are cannibalistic on eggs and pupae. However, when CS or CF are offered eggs of both species they cannot identify their own from the other species eggs (Ho, 1963), and it is not likely that cannibalism of eggs is differential with respect to genotype.

2. *Tribolium* are hosts to a number of protozoan parasites, for example, *Adelina tribolii* and *Triboliocystis garnhami*. Presence of large numbers of unhealthy looking larvae and adults and large numbers of dead larvae or pupae (dark when freshly dead) in cultures are symptomatic evidences of disease. These cultures should be discarded and the medium and all equipment sterilized. If the need arises, these cultures can be rescued and made parasite-free by the methods outlined by Stanley (1961a,b, 1964a), Bender and Doll (1963), or the simpler method suggested by the writer (Sokoloff, 1962c), which is essentially a process of parasite dilution, entailing the transfer of imagoes successively to fresh medium every 2 days, discarding the parents after the fifth transfer, and starting the new cultures with the last batch of eggs laid by the beetles.

3. A continuous watch must be maintained for the presence of mites which can propagate rapidly in beetle cultures.

4. All *Tribolium* species examined so far, and other tenebrionid beetles, secrete quinones through the odoriferous glands. This material and other metabolic wastes accumulate gradually in the medium, after which the medium is said to be "conditioned." If beetles are confined in a vessel without food, they may become excited, releasing large amounts of the highly volatile quinones. This is true more often of CF than of CS. If the presence of high concentrations of these gases does not result in the death of the adults, it may produce all sorts of abnormalities if together with the adults there are beetles in critical periods of preimaginal stages. In some instances people working with *Tribolium* may develop

an allergic reaction to the quinones released by the beetles (Park, 1934a).

5. Beetles have very small chromosomes, making cytogenetic studies difficult.

The main purpose of this review is to bring together, in easily accessible form, information available on the genetics of flour beetles and other Coleoptera, to place the material in its proper perspective, and to render clear the potentialities of these organisms for genetic research and teaching. Some of the information covered has been published in various journals, but many of the descriptions of mutants have been cited only in the *Tribolium Information Bulletin,* which is not readily accessible to many investigators.

Preserved mutant material cited in this review, insofar as possible, has been deposited at the Chicago Natural History Museum, Chicago, Illinois; American Museum of Natural History, New York, New York; and United States National Museum, Washington, D.C.

II. Taxonomic Position of Tribolium and Related Forms

Incorporated in the family Tenebrionidae are two subfamilies which include all the pests of economic importance with which this paper is primarily concerned: *Gnathocerus cornutus*, *Latheticus oryzae*, and *Tribolium* species are assigned to the Ulominae, and *Tenebrio molitor* to the Tenebrioninae.

Hinton (1948) has reviewed the taxonomic situation of *Tribolium*. Prior to Hinton's contribution, nine species had been described. He was able to include these as well as seventeen new species, one subspecies, and two forms new to the genus in five distinct species groups on the basis of morphological resemblance. Since his revision, only one new

TABLE 1
Known Species of *Tribolium*

I. *brevicornis* species group	III. *alcine* species group
A. *T. brevicornis* (Lec.)	A. *T. alcine* Hinton
B. *T. linsleyi* Hinton	B. *T. dolon* Hinton
C. *T. parallelus* (Casey)	C. *T. ceto* Hinton
D. *T. gebieni* Uyttenb.	IV. *castaneum* species group
E. *T. carinatum* Hinton	A. *T. castaneum* Herbst
F. *T. carinatum dubium* Hinton	B. *T. madens* (Charp.)
II. *confusum* species group	C. *T. freemani* Hinton
A. *T. confusum* Duval	D. *T. waterhousi* Hinton
B. *T. anaphe* Hinton	E. *T. parki* Hinton
C. *T. destructor* Uyttenb.	F. *T. cylindricum* Hinton
D. *T. giganteum* Hinton	G. *T. politum* Hinton
E. *T. downesi* Hinton	H. *T. apiculum* Neboiss
F. *T. semele* Hinton	V. *myrmecophilum* species group
G. *T. sulmo* Hinton	A. *T. myrmecophilum* Lea
H. *T. indicum* Blair	B. *T. antennatum* Hinton
T. indicum f. *seres* Hinton	
T. indicum f. *ares* Hinton	
I. *T. thusa* Hinton	

species, *Tribolium apiculum* (Neboiss, 1962) has been described. Hinton's summary, with this single addition, is included in Table 1.

On the basis of morphological resemblance and geographic distribution of the various species groups. Hinton suggests that the genus *Tribolium* must have originated from a common ancestor in the late Cretaceous or possibly earlier.

III. Cytology of Tribolium Species

Among studies which have been concerned with the cytology of Coleoptera, those of Smith (1949, 1950, 1951, 1952a,b, 1953, 1958, 1959a,b, 1960) have been particularly extensive. His examinations of beetles in 191 species, 127 genera, 66 families, and 51 superfamilies have led Smith (1950) to conclude that the primitive number of chromosomes in the Coleoptera is nine pairs of autosomes, an X about the size of the autosomes, and a minute y, "both being V-shaped and associated during maturation divisions at two terminal contact points in the form of a 'parachute.'" This formula is characteristic of the Tenebrionoidea as a whole, but a number of species in this superfamily have fewer chromosomes. These are considered to be derived species. Within the genus *Tribolium*, *T. confusum* and *T. destructor* have a much larger pair of X- and Y- (neo-X and neo-Y) chromosomes, and eight pairs of autosomes, whereas *T. castaneum* has nine pairs of autosomes, an X about the size of the autosomes, and a small y, these two appearing in the form of a parachute, conforming with the primitive condition (for illustrations and a fuller account of the following brief summary see Smith, 1952a,b). *Tribolium madens* also has this primitive formula, except that this species, in addition, has a small pair of supernumeraries (Smith, 1960b).

Tribolium confusum and *T. destructor* "descended from a form with a chromosome constitution like *T. castaneum* following fusion of the X chromosome with one pair of autosomes, the homologue of which took over the mechanical functions of the minute y chromosome." This conclusion is based on the staining reactions of the sex chromosomes and autosomes in these three species of *Tribolium* when stained with Feulgen's leucobasic fuchsin in the various gonial stages:

1. *Spermatogonial metaphase.* In *T. confusum* spermatogonial metaphases two chromosomes are occasionally "split" precociously; the larger (J-shaped) is the largest of the chromosome complement and has no morphological equivalent, its longer arm being usually visibly split into its two component chromatids. The other member is probably telocentric, it is visibly split, and is identical morphologically to the longer arm of

the J-shaped chromosome. In oogonia of this species there are two chromosomes corresponding to the J-shaped one in the male, and none corresponding to the telocentric ones. Hence, the J-shaped chromosomes in both sexes must be the X and the telocentric one the Y.

In *T. destructor*, spermatogonial metaphases exhibit one large chromosome resembling the X in *T. confusum*. The Y is much smaller, spherical and considerably broader than the minute in *T. castaneum*, its diameter being roughly equivalent to the cross section of an autosome.

2. *Pachytene.* In *T. castaneum* the X and y are condensed, deeply staining, and associated with the nucleolus.

In *T. confusum* (and probably *T. destructor*) the X and Y are also associated with the nucleolus. The bivalent consists of a short, condensed, heavily stained segment and a long, diffusely stained portion indistinguishable from the ordinary euchromatic elements.

3. *Diakinesis.* In *T. castaneum* all the bivalents except the Xy stain with equal intensity; at first the latter bivalent is more deeply stained, but later (by prometaphase) it does not differ from the others in staining reaction.

In *T. confusum* the XY bivalent differs from the other bivalents in stainability: the segment with a cross section equivalent to that of the autosomes stains as deeply as the autosomes, but the other longer, thinner segment is for the most part lightly stained, except that it terminates in a deeply stained knob.

4. *Metaphase.* In *T. castaneum* the Xy_p bivalent stains perhaps somewhat less deeply than the autosomes. This bivalent lies, at first, just off the equatorial plate, although always oriented on the spindle with the y nearer the plate and the X more distal.

In *T. destructor* the X and Y bivalent form a striking heteromorphic pair which stains with the same intensity as the autosomal bivalents, and fails to show the delayed coordination characteristic of the Xy_p bivalent of *T. castaneum*.

In *T. confusum* the XY bivalent has three major components in regard to staining reaction to Feulgen: "the differential arm of the X—positively heteropycnotic at pachytene but indistinguishable from the autosomes at metaphase; the pairing arm of the X—euchromatic at pachytene and also at metaphase; and the Y chromosome—indistinguishable from euchromatin at pachytene but negatively heteropycnotic at metaphase." Smith (1952b) believes that of the three components of the sex chromosomes, it is those of equal length, the telocentric Y and the longer arm of the X, which are the pairing arms which unite terminally at the first spermatocyte metaphase. "The remaining component, the shorter arm of the X, must be the differential arm: a relic of the X-chromosome of an Xy_p sex-determining system."

According to Smith (1952a) therefore, in *T. confusum* "the hetero-pycnosis of the Y has been developed at the expense of euchromatin during the phylogenetic history of the species." He believes that hetero-pycnosis largely denotes that the neo-Y has become genically inert. This is to be inferred from the size and stainability of the neo-Y in *T. destructor*. In this species the heteropycnotic segment, so conspicuous in *T. confusum*, is absent. Thus, since the heteropycnotic segment has been elimi-nated, *T. destructor* must have evolved from *T. confusum*, and the genes in this segment "were already more or less inert at the time of their loss."

As Smith (1952a) has stressed, the numerical relationship of the sex-determining genes to sex has remained unaltered, but with the trans-located autosome forming the neo-X, it is expected that the number of fully sex-linked genes in *T. confusum* and *T. destructor* must be consider-ably increased relative to the number of sex-linked genes in *T. castaneum*.

IV. Mutants in Tribolium

A. Tribolium castaneum

1. Sex-Linked Genes

a. VISIBLES

(1) Paddle (*pd*, Park and Frank, 1951).* A sex-influenced recessive gene of variable expression but complete penetrance and good viability. It is the anchor gene for the X-chromosome, all early sex-linked data having been reported to the left or right of *pd*. In the female the effect (with few exceptions which may resemble the male in expression) is confined to the antennal club, which may be variously fused (Plate 1,D). Usually fusion involves segments 9–10 or 10–11, but both antennae need to be examined since one may exhibit fusions and the other may be free of them. In the male the club segments are fused into a solid mass resembling a paddle; in addition, there is a loss of one or more funicular segments (Plate 1,C).

Park and Frank (1951) apparently were unaware that *pd* has effects on structures other than the antennae. The writer has recently found that the tarsus (primarily of the middle legs, but the tarsi of other legs may show the same defect) is variously modified: the penultimate tarsomere may be reduced in size and displaced off the longitudinal axis, lying more or less under the preceding tarsomere; or the last two tarsomeres may be fused with few hairs remaining as evidence of segmentation; or the segment may be completely missing. More proximal tarsomeres may also be affected (Plate 1,C). As in the case of the antennae, the effect on the tarsi is sex-influenced, being more frequent and pronounced in males. In a sample of the original *pd* strain recently provided by Dr. Thomas Park, University of Chicago, 83/243 males (34.1%) and 4/126 females (3.2%) exhibited tarsal fusions, malformations, or elimination of segments. A sample of the abnormalities observed in the males can be seen in Table 2.

* Name and date following a gene symbol are discoverer and date of discovery and not a "name and date" reference citation.

The four females found to have abnormal tarsi had these abnormalities:

(*a*) Fourth tarsal segment of second right leg and third of left hind leg reduced.

(*b*) Fusions of segments 3–4 of right hind leg; fourth of middle left reduced.

(*c*) Symmetrical reduction of fourth tarsal segment of both middle legs.

(*d*) Right legs normal; fourth segment of left foreleg and third segment of left hind leg reduced.

The *pd* gene also affects productivity: it has been found to be more productive (Phillips and McDonald, 1958) or less productive (Bartlett

TABLE 2
Tarsal Abnormalities Observed in Paddle Males*

Number	Right leg			Left leg		
	I	II	III	I	II	III
1	—	4s	—	—	1–2	3s
2	—	—	—	—	4s	3s
3	—	4s	3s	—	4–5	3s
4	—	—	3s	—	1–2, 4s	3s
5	—	—	3s	—	4s	2–3
6	—	4s	—	—	4s	3s
7	4s	1–2, 4s	3s	—	4s	3s
8	4s	1–2, 4s	2–3	—	4s	2–3
9	4s	4s	2–3	4s	4s	2–3
10	4s	4s	—	—	4s	3–4
11	—	—	—	—	—	3s
12	—	4s	3s	—	4s	3s
13	—	4s	3s	—	4s	3s
14	4s	4s	2s, 3–4	4s	1–2, 4s	3s
15	4s	1–2, 4s	3s	4s	4s	3s
16	4s	4s	3s	4s	4s	3s
17	4s	4s	—	—	4s	3s
18	4s	4s	3s	4s	4s	3s
19	—	4s	3s	—	4s	3s
20	—	4s	—	—	4s	—

* The numbers indicate tarsomeres numbered from proximal to claw-bearing segment. Numbers joined by a dash indicate a fusion of those segments. The letter "s" indicates a tarsus reduced in size.

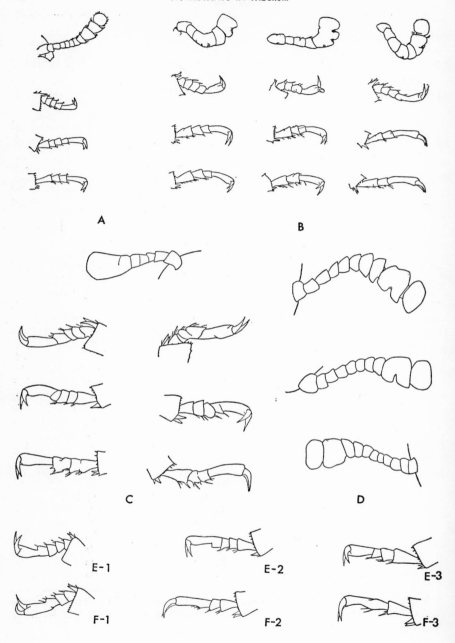

A

B

C

D

E-1

E-2

E-3

F-1

F-2

F-3

et al., 1962b) than wild type, apparently as a result of the residual genotype (Sokoloff *et al.,* 1965).

(2) Miniature appendaged (*ma*, Sokoloff, 1959; *ma*-1, Sokoloff,
1964). This semilethal recessive mutant has a stunted appearance. It
is recognizable as soon as the pupa forms by the shape of the head,
length of elytra and membranous wings, and the position and shape of
the legs. In the imago, the elytra cover only about two-thirds of the
abdomen; the membranous wings often lie extended over the abdomen
or may droop at the sides; the head, thorax, and abdomen are more
compact; the antennameres also appear reduced in size (Plate 2,B). On
the ventral surface, the podomeres of all legs (with the possible exception of the coxa) are markedly modified, being shorter and thicker than
in the normal legs (Plate 2,D). The *ma* gene is located about 12 units
to the left of *pd,* about 2 units to the left of pygmy (*py*) (Sokoloff,
1960d).

(3) Red (*r*, Lasley, 1960). A recessive of excellent viability, variable
expression, but complete penetrance, located about one unit to the left
of *pd* (Sokoloff *et al.,* 1960a). The normally black pigment is eliminated
from the ommatidia, but this mutant, like all other eye color mutants,
appears bicolored or "spectacled" owing to the presence of melanin pigment in the ocular diaphragm, an endoskeletal structure serving to support the eye. The central part of the compound eye, lying over colorless
tissue within the ocular foramen, consists of ommatidia which appeared
pink to Bordeaux red in the original stock. In time, the eye pigment
of this mutation gradually faded to a much lighter shade, including some
beetles without detectable reddish pigment (resembling pearl, *p*). Since
on outcrossing the darker colors are restored, the presence of modifiers
is suggested.

The *r* gene has pleiotropic effects: the Malpighian tubules, normally
black or dark brown, become pale or transparent in the *r* mutant (Shaw,
1965).

A number of alleles or possible recurrences of *r* has been found: Daw-

---

PLATE 1. Antennal and tarsal mutations in *Tribolium castaneum*. (A) Normal
antenna and tarsi. (B) Antennae and tarsi of three serrate (*ser*) males showing
variation in fusion of antennameres and tarsomeres. (C) Left antenna and tarsi of
three pairs of legs from the same paddle (*pd*) male showing variation in fusion of
tarsomeres. (D) Three antennae of *pd* females showing partial or complete fusion
of segments 9–10, (E1–3, F1–3) Right tarsi of fore-, middle, and hind legs of the
fas-3 mutant.

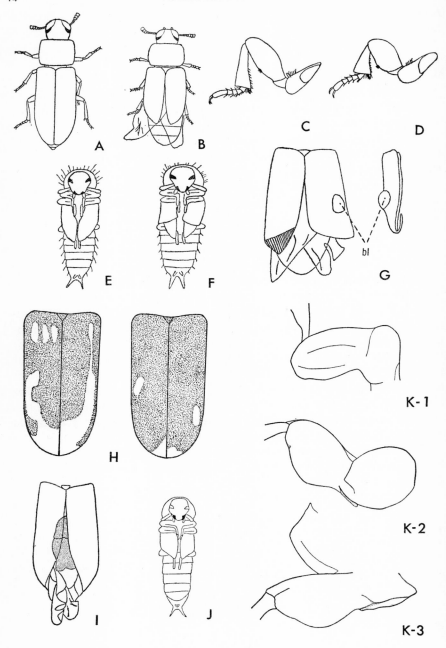

son found r^D which differs from r and other alleles in being sex-influenced: in males the eye is light red and easily distinguishable from wild type; in females the mutant is easily distinguishable from normal beetles in the pupa or young imagoes (Dawson, 1964a) but in aged females it can be distinguished from the wild type only by strong illumination and by casting a shadow over the eye (Sokoloff, 1964, unpublished). Ho found r^H in the F_1 of a sample derived from an Oakland, California, cereal mill (Ho and Sokoloff, 1962). Eddleman and Bell (1963) have reported *r-1* as producing a darker red color than *r*, but these differences would probably be nullified if *r* and *r-1* were to be introduced in the same residual background.

Histological preparations reveal yellow and pink granules in moderate amount in the ommatidia of *r-1*, while in the normal eye the granules are in extremely heavy concentration and blackberry-red in appearance. (Eddleman and Bell, 1963).

The fact that so many independent discoveries of *r* or its alleles have been made in unrelated strains of *T. castaneum* suggests that r^+ is unstable, mutating at a high frequency to *r* or other alleles. Evidence that mutation may also occur in the reverse direction at least in somatic tissues is derived from the discovery by Dawson (1963b) of a male with one eye red, the other eye red on its dorsal aspect (above the gena) and black on its ventral aspect. Although the beetle died before it could be tested, it was probably a red hemizygote.

(4) Pygmy (*py*, Lasley, 1960) is a semidominant affecting body size, located about 2 units to the left of *ma* (Sokoloff, *et al.*, 1960a). Fecundity in females is greatly reduced; viability is variably reduced in various crosses: in some cases not at all, in others 20%, in still others more than that, but there is no difficulty in maintaining it as a stock. The length of the body is reduced to about two-thirds of the normal, while pupa weight, as shown in the following tabular material, is reduced to about half in *py* and less than that in $+/py$.

PLATE 2. Sex-linked mutants in *Tribolium castaneum*. (A) Dorsal view of normal beetle. (B) Dorsal view of miniature appendaged (*ma*). (C) First leg of normal beetle. (D) First leg of *ma*. (E) Normal pupa. (F) Truncated elytra (*te*) pupa. (G) Dorsal and lateral views of *te* adult. In the figure "*bl*" represents a blister often found on the elytra of *te* beetles. (H) Two examples of elytral pigmentation in the spotted (*sp*) mutant. (I) Appearance of divergent elytra (*dve*) teneral adult. (J) *dve* pupa. (K1–3) Right fore-, middle, and hind leg of deformed podomeres (*dpm*). (Drawings A–J, by permission of the Editor, *Canadian Journal of Genetics and Cytology*.)

Pupa Weight in Milligrams*

| | ♂♂ | | | | ♀♀ | | | | |
| | +/ | | py/ | | +/+ | | +/py | | py/py | |
Mating	No.	Wt.	No.	Wt.	No.	Wt.	No.	Wt.	No.	Wt.
1. +/py × py/	26	2.732	20	1.326	—	—	27	2.510	20	1.362
2. +/py × py/	29	2.706	34	1.397	—	—	31	2.555	25	1.417
3. +/py × +/	21	2.583	19	1.168	13	2.765	17	2.445	—	—
4. +/py × +/	5	2.742	15	1.357	21	2.824	14	2.524	—	—
5. +/py × +/	27	2.564	18	1.163	20	2.746	11	2.476	—	—
6. +/py × +/	16	2.669	25	1.281	16	2.805	15	2.487	—	—

* Data from Lasley (1960c).

The above data indicate a clear-cut difference between wild-type males and females in the Chicago strain, which amounted to 0.145 ± 0.45 mg and which did not differ significantly from the difference of 0.128 mg in the two sexes based on thousands of male and female wild-type pupae. The average difference between normal and heterozygous pygmy (+/py) females, 0.302 ± 0.043 mg, suggests that the py gene affects the weight of the carriers. The py gene is actually a semidominant, but heterozygotes can be identified only by weighing them; therefore, in practice it is easier to consider it as a recessive.

(5) Divergent elytra (dve, Sokoloff, 1961) is an incompletely recessive gene. In the pupa it is identifiable by the more pointed appearance of the elytra, and membranous wings. The wings are directed away from the abdomen, appearing shorter in camera lucida drawings, exposing the tarsi of the hind legs (Plate 2,J). In the imago the elytra are more pointed and divergent, the divergence starting sometimes at the scutellum but more often at some distance away from this structure. Sometimes a blister may develop on one or both of the membranous wings (Plate 2,I), but the fluid is withdrawn into the body cavity before sclerotization is completed. Imagoes may die at an early age if low humidity prevails. Viability of dve is reduced by about 20%. This gene is located about 11 units to the left of py (Sokoloff, 1963f).

(6) Truncated elytra (te, Sokoloff, 1960). An incompletely recessive gene (detectable in a small proportion of +/te females), having lethal to semilethal effects. In the te pupa the elytra may appear cut off about the middle, exposing the membranous wings (Plate 2,F) or they may appear curved and drawn completely away from the hind wings. In the

adult the expression is also variable: both elytra may appear truncated (but viewed from the side it is evident that the tips are only folded under the proximal portions), or the distal ends may appear bent toward the abdomen. Beetles combining these two phenotypes and developing a permanent blister on the elytra may also be found (Plate 2,G). The *te* gene has lethal to semilethal effects distorting the recombination values in three-point crosses: two separate studies (Sokoloff, 1960, 1962e) located *te* 12–24 units to the right of *pd*. Dawson (1963a) found *te^D* a less lethal allele of *te*. He located *te^D* about 11 units to the right of *pd*, and determined that mortality (at least in males) occurs predominantly in the pupa stage, but a fairly large number of beetles die when the imagoes fail to free themselves of the pupal skin.

(7) Spotted (*sp*, Sokoloff, 1960) is an incompletely recessive mutation affecting the pigmentation of the elytra. Normally *T. castaneum* has a uniform red-rust or chestnut body color. In *sp* unpigmented areas appear, chiefly detectable in freshly eclosed imagoes. These areas may be restricted to the tips of the elytra or they may extend as more or less symmetrical stripes almost to the anterior margins of the elytra (Plate 2,H). On aging these light areas disappear (becoming pigmented to about the same degree as the rest of the body) and *sp* cannot be distinguished from non-*sp* beetles. This gene is located to the left and about 46 units from *pd* (Sokoloff *et al.*, 1960a; Sokoloff, 1962e). Although linkage tests involving this gene require careful planning of the experiment to identify the *sp* and non-*sp* beetles as they emerge, the extra labor involved is outweighed by the ease in classification of *sp*, its viability, and the relatively large distances between this gene and other useful markers such as *py*, *r*, and *pd*.

(8) Red modifier (*M^r*, Sokoloff, 1961). This gene is located about 17 units to the right of *r* (beyond *te*). It is unusual in that it behaves as a modifier of red only in hemizygotes. In females homozygous for *r* but heterozygous for the modifier (*rM^r/r^+*) it behaves as a suppressor (Sokoloff, 1965). The situation is so peculiar that it will be discussed at more length in a later section (see Chapter VI).

(9) Pokey (*pok*, Dawson, 1962) is a recessive semilethal superficially resembling, and epistatic to, *dve*: the elytra fail to meet over the dorsal surface of the abdomen, leaving one-fifth to one-third of the abdominal surface exposed (see Dawson, 1964c for illustration). In addition, the antennal segments may be fused and the legs may appear shorter and thicker than normal. Mutants develop extremely slowly. The location of *pok* is about 2 units to the left of *dve* (Dawson, 1964c).

(10) Serrate (*ser,* Dawson, 1963). A recessive gene affecting the antennae and tarsi unequally in the two sexes. In a survey of fifty females over 50% showed antennal fusions in the funicle or in the club, 18% showed fusions in both the funicle *and* club, and 30% showed no antennal fusions whatsoever (both antennae having been scored). Of fifty males examined, one male showed no antennal fusions; 80% showed fusions of the club *and* funicle (Plate 1,B), and the remaining 14% were about equally divided: 6% showed fusions of the funicle, 8% of the club. The same females, scored for tarsal fusions, showed a reduction of fusion of tarsi in at least one of the middle legs in at least 56% of the cases; 30% showed reductions or fusions of the middle and hind legs; the remaining 14% had reductions in tarsomeres of at least one member of the front and middle legs, but only in two females did the effect extend to the hind legs. In the males, almost half had a reduction in tarsomeres of at least one member of all three pairs of legs (Plate 1,B). In the remaining beetles, the tarsus of the first pair of legs was not affected, but the tarsi of the middle legs were reduced by one or two segments; in only nine out of the fifty males examined there were no visible deformities of the hind legs. Thus, for both sexes, the tarsi of the middle legs provide the best criterion for classification of *ser.* Tables 3 and 3a

TABLE 3

Fusions in the Serrate Mutant*

	Males		Females	
Number	Right	Left	Right	Left
1	4–5, 10–11	4–5, 10–11	4–5, 9–11	4–5, 9–10
2	4–5, 6–7, 8–9, 10–11	4–5, 6–7, 8–9	10–11	9–10
3	4–5	4–5, 6–7	0	0
4	9–10	9–10	0	0
5	4–5, 10–11	6–7, 10–11	8–9	0
6	4–5, 6–7, 10–11	4–5, 9–11	0	0
7	4–7	4–8, 9–11	0	0
8	3–5, 9–10	3–5, 9–10	0	10–11
9	3–7, 8–9	4–5, 6–7, 9–10	9–11	10–11
10	4–5	4–5	10–11	10–11

* Numbers represent the antennameres numbered from proximal to distal. Fusions of adjoining segments are represented by a dash. Where more than two segments are fused to each other the dash connects the proximal and the most distal antennamere of the fused block.

TABLE 3a

Numbers of Tarsomeres in 10 Males and 10 Females Derived from the Serrate Stock*

	Males						Females					
	Leg 1		Leg 2		Leg 3		Leg 1		Leg 2		Leg 3	
Number	R	L	R	L	R	L	R	L	R	L	R	L
1	5	5	4	3.5	4	3.5	4.5	5	4	4	3.5	4
2	5	5	4	4	4	4	5	5	4	4	4	4
3	5	5	4	4	4	4	5	5	4.5	4	4	3.5
4	4	4	3.5	3.5	3	3.5	5	5	4.5	4	4	3.5
5	4.5	4.5	4.5	4	3.5	4	5	5	4	4.5	4	4
6	5	5	3.5	3.5	3.5	3.5	5	5	4	4	4	4
7	4	4	3	3	3	3	5	5	4	4	4	3.5
8	5	5	4	4	4	4	5	5	4	4	3.5	4
9	4.5	5	3.5	3.5	3.5	3.5	4.5	4.5	4	4	3.5	4
10	5	5	4	4	3.5	3.5	5	5	4	3.5	4	4

* A figure followed by a decimal indicates partial fusion of adjacent tarsomeres.

show the variation in antennal and tarsal fusions in a sample of ten male and ten female *ser* beetles.

The *ser* gene is not allelic with *pd,* but its location is very near *pd:* it is located less than one unit to the right of *r* (Dawson, 1965).

(11) Deformed podomeres (*dpm,* Sokoloff, 1964). Recessive of incomplete penetrance and variable expression found in linkage studies between *ser, py,* and *pd* (Sokoloff, 1965c).

In strongly expressed phenotypes the femur is nearly globose, the trochanter is missing, and the proximal end of the femur is fused to the coxa, but the tibia and tarsi appear not to be affected to any large extent (Plate 2,K). In mildly deformed beetles the femur of only one leg, most often the mesothoracic one, less often the femur or the first pair of legs, is short and thick but the remaining legs have normal femora. Occasionally the tibiae are bent as in *btt,* and they may, in fact, be deformed because of this autosomal gene. Preliminary counts suggest the position of *dpm* to the right of *pd.* The exact location of *dpm* will be difficult to establish, however, since penetrance is very low.

The group of investigators at Purdue University, Lafayette, Indiana, has contributed the following limited information on three sex-linked visibles.

(12) Ring eye (*rg*, Yamada, 1961), a recessive eye color gene resembling pearl (*q.v.*) in that the lighter colored center of the compound eye is circled by a darker marginal area. Good penetrance and expressivity and normal viability (Yamada, 1962). No information on its location in the X-chromosome.

(13) Tarsal irregular (*ti*, Shideler, 1960), a recessive with variable expression: the number of tarsal segments for the second pair of legs is reduced from five to four by an elimination of any of the first four tarsal segments or by a fusion of the third and fourth tarsal segments. Frequently the pro- and metathoracic legs may also be affected (Shideler, 1962). Penetrance is improved if the mutant is cultured at a high temperature: at room temperature (24.4°C) 71% penetrance was observed; at 32.8°C penetrance was enhanced to about 97% regardless of whether the beetles were reared at 40 or 70% relative humidity. At 37.8°C and 70% humidity penetrance was complete (Krause, 1963b).

(14) Rose (*rs*, Reynolds, 1964). Spontaneous. A recessive with complete penetrance, good expressivity, and excellent viability. Phenotype in the pupa and young imago resembles red, but in old adults it darkens toward wild type. Located 15 recombination units to the left of *py* (Reynolds, 1964). On the basis of these results it is probable that *rs* is located to the left of *dve* (since this gene is located 11 units to the left of *py*), and probably to the left of *pok*, but three-point crosses would be desirable.

b. Sex-Linked Lethals

Four sex-linked lethals have been mapped: lethal-1 (*l₁*, Sokoloff, 1961) and lethal-4 (*l₄*, Sokoloff, 1961) were found in the F_1 descendants of adult females emerging from eggs rocketed 55 miles into space December 4, 1959, and tested for the presence of lethals; lethal-2 (*l₂*, Dawson, 1962) was found in the *Sa-2* stock; lethal-3 (*l₃*, Sokoloff, 1961) was found during linkage tests determining the position of *dve* in respect to *r* and *py* and it probably originated in the *py* stock.

The position of *l₁* is 39 units to the right of *pd*; *l₃* is located about 5 units to the left of *py*, between *dve* and *py*; *l₂* and *l₄* appear to be allelic, and they are located 31–33 units to the left of *pd*, which places them between *dve* and *sp* (Sokoloff and Dawson, 1963b). An unexpected finding while determining the location of *l₂* and *l₃* was a preponderance of females heterozygous for these lethals (for example, all seventeen F_1 females of crosses between $+/l₃ \times dve\ pd$ chosen to produce more progeny were heterozygous for the lethal). The available data from linkage experiments could not resolve whether this phenomenon was the result of meiotic drive or the superficially similar phenomena of gametic competition or gametic selection (Sokoloff and Dawson, 1963b). However, this phenomenon is not unique for *T. castaneum* as will be pointed out later in certain studies of sex-linked genes in *T. confusum*.

Dawson, while attempting to map the l_2 lethal found that F_1 females and their female progenies gave aberrant ratios: 3 females:1 male instead of the expected 2:1 ratio (Dawson, 1962c; Sokoloff and Dawson, 1963b). This phenomenon also requires further investigation.

2. Autosomal Genes in Established Linkage Groups

a. Linkage Group II

(1) Pearl (*p*, Park, 1937), a recessive affecting the pigment of the eye, was the first mutation described for *T. castaneum* (Park, 1937). The normal eye appears black in the adult and the ocelli in the larva also are black. In *p* larvae of any instar no ocelli are detectable, but this is probably because the lack of pigment renders these structures invisible. The pupal and adult eye also are devoid of pigment except that the marginal ommatidia appear black because they form over the ocular diaphragm, an endoskeletal structure pigmented black. The central ommatidia, placed over the ocular foramen, appear crystalline (Plate 3,E,J). The black pigment in the ocular diaphragm begins to develop in the late pupa, hence the eye already appears bicolored before imagoes eclose. The mutation pearl has good viability, and complete penetrance, and is a useful marker for the identification of this linkage group. Shaw (1965) has found that *p* has pleiotropic effects: the normally black or dark brown Malpighian tubules become pale or transparent in the *p* adult.

The wild-type allelomorph of pearl appears to be unstable: somatic mutations in beetles heterozygous for pearl and resulting in imagoes having one eye black, the other completely pearl; one eye black, the other half pearl; or both eyes with small pearl areas, occur at a frequency of about 1:10,000 (Sokoloff, 1959; Sokoloff and Shrode, 1960).

Waddington and Perry (1963) have examined the eye in *Tribolium castaneum* with the aid of the electron microscope. Their preliminary observations show that the ultrastructure of the various cell types is essentially similar to that of the Dipteran eye. "The rhabdom, composed of a tightly packed array of hexagonal tubules, is cup-shaped and is situated in the distal region only of the retinula cell group. The

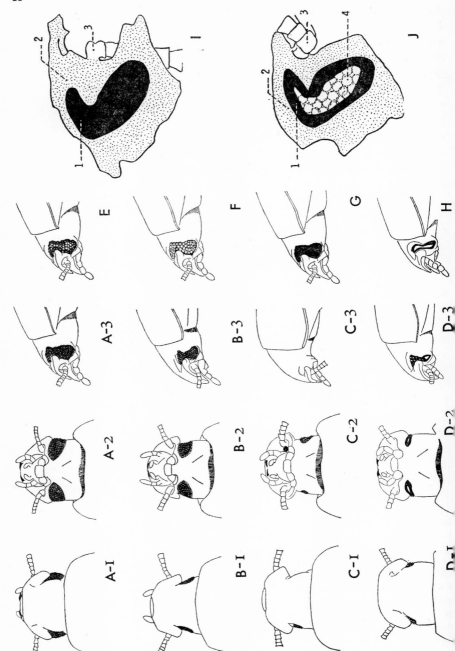

rhabdomeres of six retinula cells are fused at the lateral edges of adjacent cells, the seventh and eighth cells lie at the base of the cup, and the eighth cell has no rhabdomere. Onset of rhabdom formation occurs during the third day of pupation, when the membranes of the retinulae at the centre of the ommatidia are thrown into shallow folds. These later deepen to form the tubules of the adult rhabdom. In contrast to the situation in the Drosophila eye, in which there is simultaneous folding of the membrane over the surface of the retinulae, where the rhabdom will eventually be formed, there seems to be a progressive change in the Tribolium eye, with the area of membrane folding gradually extending outwards from a central point, during mid-pupal life."

(2) Pink (p^{Pk}, Lasley, 1960) is recessive to wild type but dominant to and allelic with pearl. In the young imagoes the two alleles may appear the same, but in p^{Pk}/p or p^{Pk}/p^{Pk} a pinkish tinge develops in the ommatidia as the beetles become older (Lasley, 1960b).

(3) Pegleg (*pg*, Lasley and Sokoloff, 1961). A recessive of good penetrance but variable expressivity affecting the legs of the preimaginal and the imaginal stages. (Plate 4, E 1–3). In the adult one or both members of any pair of legs may be affected in the following ways: (a) the tarsi may be completely missing, or the onychium, bearing claws, may be attached to the tibia; (b) all tarsi and part of the tibia may be missing, and the remaining portion of the tibiae may taper toward the distal end; (c) all proximal tarsal segments may be missing, the tibia is badly deformed (short and thick), and part of the onychium develops on the lateral aspect of the distal end of the femur; (d) in the extreme expression of *pg* podomeres of all the legs except the coxa may be missing, and yet the beetles manage to eclose from the pupa; (e) in what is regarded as the mildest expression of the gene all the podomeres are present, but the femur is short and curved and the tibia badly twisted, resembling *dfl* (*q.v.*). Location of *pg* is about 30 units from pearl. How-

---

PLATE 3. Eye mutants in *Tribolium castaneum*. (A1–3). Dorsal, ventral, and lateral views of normal beetle. (B1–3). Dorsal, ventral, and lateral views of Microphthalmic (*Mo*). (C1–3) Dorsal, ventral, and lateral views of microcephalic (*mc*). (D1–3) Dorsal, ventral, and lateral views of Bar eye (*Be*). (E) Lateral view of pearl (*p*). (F) Lateral view of *p* with light ocular diaphragm (*lod*). (G) Lateral view of glass (*gl*). (H) Lateral view of squint (*sq*). (I) Medial view of the dissected eye in the normal beetle. (J) Medial view of the dissected pearl eye, 1, ocular diaphragm; 2, inner surface of head exoskeleton; 3, antenna; and 4, outline of eye facet. (Drawings I,J by permission of the Publisher, *American Naturalist;* A–F and H by permission of the Editor, *Canadian Journal of Genetics and Cytology.*)

ever, this is only an estimate, since the values were derived from a cross showing a deficiency in the *pg* classes (Lasley and Sokoloff, 1961).

b. LINKAGE GROUP III

This linkage group is identified in *T. castaneum* by the semidominant body color gene black (*b*) and the recessive light ocular diaphragm (*lod*).

(1) Black (*b*, Sokoloff, Slatis, and Stanley, 1960) occurred spontaneously at about the same time in two geographically separated, unrelated wild-type strains, one designated Chicago and the other McGill. Tests of allelism indicated that these were independent occurrences of the same gene (Sokoloff *et al.*, 1960b), and for simplicity they may be referred to as *b* and *b-1*, respectively. The normal body color of *T. castaneum* is red-rust or chestnut. The *b* gene causes the formation of a black pigment, which under the dissecting microscope appears to match 65'UR-Vm (dull purple), 69' RV-Rm (aniline black), or 61' VR-Vm (dull violet) on Plate L, or possibly 65' RR-Vm (raisin black) on Plate XLIV of Ridgway (1912), "Color Standards and Color Nomenclature." The heterozygote is nearer wild type than black and it is referred to as bronze. It is possible to distinguish *b/b* from non-*b* larvae in the first instar by examining the head, tergites, and urogomphi. It is also possible, with some training, to distinguish *b/+* from +/+ larvae if they are, say, middle-sized, by examining the urogomphi. Owing to the fact that the urogomphi are very narrow in the pupa, it is not possible to distinguish +/+ from *b/+* with certainty, but these nonblacks can be separated readily from *b/b*.

Several possible recurrences of black have been observed: the writer found: b^s (Sokoloff, 1962d) in the F_2 of adults emerging from eggs sent 55 miles into space (Sokoloff, 1962d); $b^s - 1$ was found as a single heterozygous female from a strain maintained by Dr. Howard Erdman at the General Electric Laboratories, Richland, Washington (Sokoloff, 1964h). These and the above-mentioned mutants were all probably allelic or recurrences of an unnamed mutant described by Miller (1944). Eddleman (1962) reported a lighter black called cordovan, whose color corresponds to color samples 8H8 (Maertz and Paul) and 7 pl (Jacobson). He has come to the conclusion that cordovan is an allele of *b* and designated it b^{cd}. Beetles genetically *b/+* and b^{cd}/b^{cd} are nearly identical in phenotype; b/b^{cd} beetles are darker than b^{cd}/b^{cd}, but lighter than *b/b* (Eddleman, 1964). Dewees (1963) found b^D, an allele of *b*, in a sample collected from a food storage sack at Southern Illinois University. Dyte (1964) found a recessive allele of *b* called tawny (b^t). Since b^{co}

has not been made available (Dyte, 1964, personal communication) tests of allelism between b^{cd} and b^t could not be made.

(2) Light ocular diaphragm (*lod*, Sokoloff, 1962) is a recessive gene of good viability and expressivity and complete penetrance blocking the deposition of black pigment (presumably melanin) in the endoskeletal structure called the ocular diaphragm (Plate 3,J). In all tenebrionid eye mutants examined, this structure is present under the marginal facets. The central ommatidia lie over the ocular foramen. When the diaphragm is pigmented, all eye color mutants have a bicolored or spectacled appearance. If these eye mutants are *lod/lod* homozygotes as well, then the ommatidia become uniformly colored throughout the eye (Plate 3,F). There are about 24 units between *lod* and *b* (Sokoloff, 1964i), but if *lod* is to be used as a tester gene, linkage tests are easier to perform if preliminary crosses are made to combine the new mutant with the same eye color gene which has been combined with *lod* to obtain a doubly homozygous tester stock. For example, in our laboratory *lod* has been combined with *p*. Combining a hypothetical mutant *m* with *p* and crossing with *lod p*, F_1 genetically *m/+*; *p/p*; *lod/+* are obtained. Some of these are stored in virginal condition until some *m/m*; *p/p*; *lod/lod* are available in the F_2. The stored F_1 are then backcrossed to the triple recessives to determine whether *m* is linked with *lod*. Dewees (1963) found an allele of *lod*, *lod*^{D}, in a "natural" population in a sack of grain stored at Southern Illinois University, Carbondale, Illinois.

c. Linkage Group iv

This linkage group is identified by a number of interesting and useful mutants.

(1) Sooty (*s*, Bartlett, Bell, and Shideler, 1960) was reported as recessive by Bartlett *et al.*, (1962a), but a few individuals can distinguish the heterozygote from the wild-type homozygote (Schlager, 1963; Sokal and Huber, 1963); hence it is really semidominant, but considerable training is required before the heterozygote can be identified with certainty. The phenotype of *s/s* is nearly identical to the bronze phenotype produced by the black heterozygote (*b/+*); that of *+/s* is described as "light rust color somewhat lighter than the reddish brown wild type" (Sokal and Huber, 1963, p. 170). Viability of *s* was reported by Bartlett *et al.* to be equal to that of wild type. Introduced as a marker in the synthetic stock manufactured by Lerner and Ho (1961), however, *s* exhibits a reduction in viability of about 20% (Sokoloff *et al.*, 1963). Considering *s* as a recessive: crossing these with *b* beetles to obtain F_1, and crossing F_1 individuals to each other to obtain F_2, Bartlett *et al.*

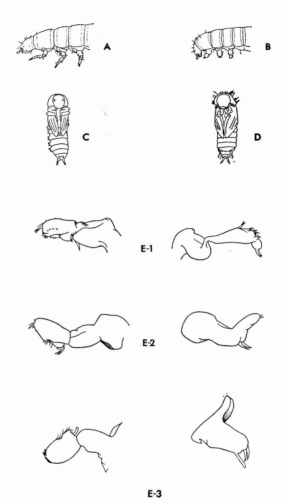

PLATE 4. Various leg mutants in *Tribolium castaneum*. (A,C) Normal larva and pupa. (B,D) Pegleg (*pg*) larva and pupa. (E1–3) The three pairs of legs in a fairly strongly expressed *pg* adult. (F,G) Larva and pupa of deformed legs (*dfl*).

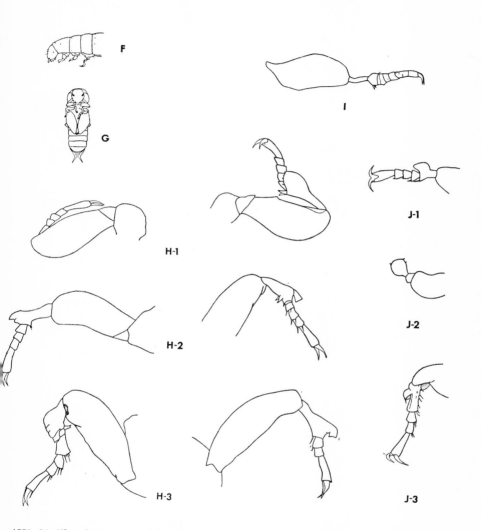

(H1–3) The three pairs of legs of adult *dfl*. (I) Single leg of *dfl-1* adult to show the narrow tibia. (J1–3) The phenotype of the *dfl-1* allele (originally called reduced tibia).

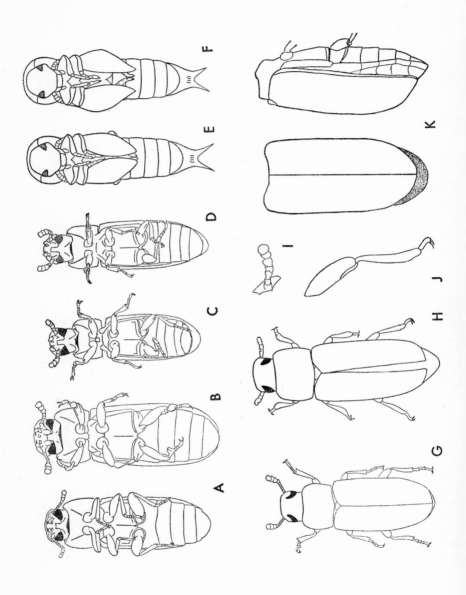

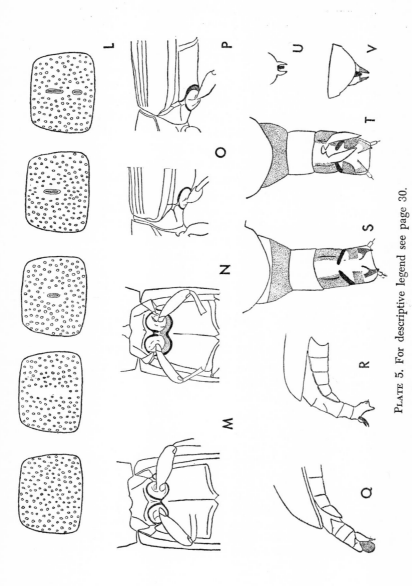

PLATE 5. For descriptive legend see page 30.

(1962a) obtained a phenotypic ratio of 4 black:2 dark-sooty; 7 sooty:3 red. These authors have pointed out that one dose of *b* produces as much pigment as two doses of *s*. The two loci act additively in the b^+/b; s/s genotype to produce more pigment than either locus does in the presence of the wild-type alleles of the other locus mutant, i.e., b^+/b; $s/s > b^+/b^+$; s/s and b^+/b; $s/s > b^+/b$; s^+/s^+. [The above ratio differs from that obtained in crosses between *b* and *j* heterozygotes. Here the array of phenotypes is 4 black:3 jet:6 bronze:3 red (Sokoloff, 1962a).]

(2) Bar eye (*Be*, Lasley and Sokoloff, 1961) is a dominant with recessive lethal effects. The eye, which in normal beetles is about six facets wide when its lateral aspect is examined, is reduced in mildly expressed *Be* to an eye one to three facets wide. The ventral portion of the *Be* eye is largely devoid of ommatidia, the ocular diaphragm lying exposed, and at times portions of this structure fail to form. A variable number of ommatidia develops on the lateral and dorsal aspects of the eye, sometimes only three to four ommatidia being identifiable (Plate 3 D, 1–3). Examination of sections of *Be* with the electron microscope by Waddington and Perry (1963) showed that in the few existing ommatidial groups there is some disorganization of the rhabdom tubules. The *Be* mutant can be identified readily in the pupa if pigment deposition has begun in the ommatidia. The lethal effect of *Be/Be* appears to occur in the egg stage. *Be* is located about 25 units away from *s*. The same recombination value is obtained in the two sexes (Sokoloff, 1962a).

(3) Deformed legs (*dfl*, Sokoloff, 1960). Autosomal recessive of poor penetrance and variable expression. Any leg may be affected, but usually only one of a given pair of legs. The tibia is the podomere most frequently affected, becoming short, curved outward or inward anteriorly

---

PLATE 5. Various mutants in *Tribolium castaneum*. (A) Ventral view of normal beetle. (B) Ventral view of Short antenna-2 (*Sa-2*). Note reduction in numbers of antennameres in the funicle and the gnarled appearance of the tibiae. (C–D) Two examples of short antenna (*sa*). The phenotype of the antennae is the same as in *Sa-2*, but of the podomeres, the femur podomere is the one most frequently affected. (E–F) Normal and akimbo pupae. (G–H) Dorsal view of normal and Fused tarsi and antennae (*Fta*) mutant. (I–J) Detail of *Fta* antenna and leg. Note reduction in number of antennameres and tarsomeres. (K) Dorsal and lateral views of short elytra (*she*). (L) Normal prothorax and four examples illustrating the range in expression of cut prothorax (*ctp*). (M–P) Ventral and lateral views of normal (M,O) and "incomplete mesosternum" (*ims*). (Q–T) Normal (Q,S) and juvenile urogomphi, *ju* (R,T) females. Note that in *ju* there is a pair of appendages, one on each side of the anal opening. (U–V) Two *ju* males. In U the dark medial mass is a fecal pellet. (A–C and H by permission of the Publisher, *American Naturalist*; K by permission of the Editor, *Canadian Journal of Genetics and Cytology*.)

or posteriorly (Plate 4, H 1–3). The mutant is detectable in the larva or in the pupa. (Plate 6, F,G). From some crosses where penetrance is improved it has been possible to determine the position of *dfl* 24 units from *s* and about 50 units from *ims*. This would place *dfl* near *Be*, but it is not possible to state at the present time whether *dfl* is to the left or right of this gene (Sokoloff, 1964e).

(4) Incomplete mesosternum (*ims*, Sokoloff, 1962) is a recessive with incomplete penetrance. The posterior medial projection of the meso-sternum ("spinasternum" of El Kifl, 1953) fails to meet and fuse with the anterior medial projection of the metasternum (mesospinasternum of El Kifl, 1953) between the coxae of the mesothoracic legs, resulting in a much freer movement of these legs (Plate 5, N,P). This gene is located about 10 units from *s* and about 43 units from *Be*, these distances being approximate owing to the incomplete penetrance of *ims*. The pre-imaginal stages have not been examined for this condition (Sokoloff, 1964e).

(5) Juvenile urogomphi (*ju*, Sokoloff, 1962) is a recessive of good penetrance and viability (up to eclosion of the imago), but variable expressivity. For lack of a better term this name was given to a mutant possessing a pair of appendages, located laterally (one on each side) to the anal opening in both sexes (Plate 5, R,T,U,V,). The appendages appear larger in females than in males, and sometimes resemble the anal cerci in Orthoptera, but there is no definite segmentation. The tips of the appendages appear sclerotized, possessing a brownish pigment. There is no evidence that these are secretory, but it has been observed that often, particularly in females, these structures are trapped by caked flour, probably after repeated defecation, with the result that more fecal matter accumulates between these appendages. Eventually the beetles are no longer capable of defecating effectively and death ensues.

So far as the writer is aware, although larvae and pupae of Coleoptera have urogomphi, these structures are absent in adults of all families of beetles. Recent comparisons between these adult structures and pupal urogomphi leave little doubt that these appendages in the *Tribolium* adult are like pupal urogomphi with minor modifications (Plates 18, 19). Hence, we have here an example of paedomorphosis (Gregory, 1946, p. 354), a situation in which the pupal characters are retained in the adult. The *ju* gene is located about 40 units from *Be* (Sokoloff, 1964e).

(6) Cut prothorax (*ctp*, Sokoloff, 1962). A recessive of good pene-trance and viability, but variable expressivity, characterized by a small, unsclerotized area in the midline of the dorsal surface of the prothorax (Plate 5,L). At times the unsclerotized area is broken up into two

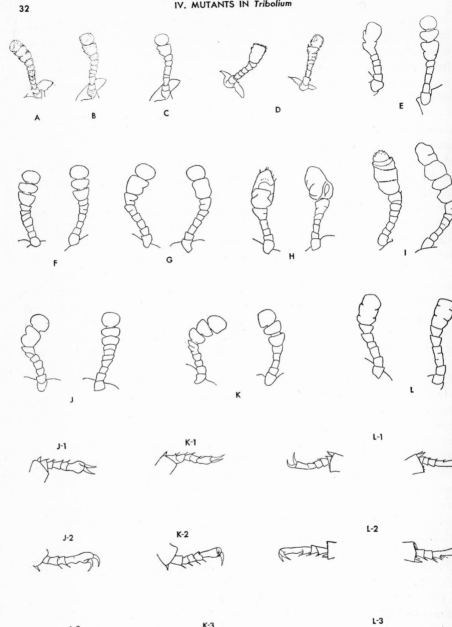

A   B   C   D   E

F   G   H   I

J   K   L

J-1   K-1   L-1

J-2   K-2   L-2

J-3   K-3   L-3

widely separated components. In some specimens the cut area "heals," sclerotizing into a smooth area devoid of the numerous pits normally present in the prothorax. The character may be detected in the larvae but only with difficulty in the pupae. The *ctp* gene was found at the same time and in the same individual as *ju*. It is located about 10 units from *ju* away from *Be* (Sokoloff, 1964e).

(7) Sternites incomplete (*sti*, Sokoloff, 1963) is a recessive of good viability and variable expression. In strongly expressed *sti*, adjacent abdominal sternites fail to overlap and one or both of the sternites fail to sclerotize, with the result that there is a wide fleshy area extending from one side of the abdomen to the other (Plate 7,N). In less extreme cases only small, unconnected, unsclerotized, and more or less symmetrically placed areas form between successive abdominal sternites. Most frequently these unsclerotized areas develop between the first and second apparent abdominal segments, but similar areas may be formed between adjacent segments located more posteriorly. The *sti* gene is located 41 units from *s* and 35 units from *ims* (Sokoloff, 1964e). It may be pointed out that a similar, apparently unconnected, abnormality develops in some mutants exhibiting a divergence of the elytra, and in some of the short elytra mutants.

(8) Fused antennal segments-2 (*fas-2*, Sokoloff, 1962). A recessive of good penetrance and viability but variable expressivity. The first individuals found were isolated because of a uni- or bilateral absence of the antennae. Beetles of subsequent generations had both antennae, but various segments, primarily of the funicle, (Plate 6,C), were fused as shown in Table 4.

It is clear from Table 4 that the modal number of adults affected by *fas-2* is in the class which includes bilateral fusions of segments 5–6 of the funicle. In addition, one female had segments 5–6 of the right antenna fused, and the left antenna bifurcated beyond the fifth segment.

PLATE 6. Antennal and tarsal mutations in *Tribolium castaneum*. (A) Normal antenna. (B) Fused antennal segments-1 (*fas-1*). (C) Fused antennal segments-2 (*fas-2*). (D) Two examples of Deformed (*Df*). (E) Two examples of Spatulate (*Spa*). (F) Right and left antenna of Fused antennal segments-4 (*Fas-4*). (G) Right and left antennae of pectinate (*pec*). (H,I) Right and left antennae of the fused antennal segments-6 (*fas-6*) beetles. (This mutant is allelic with *ap*^{D} and has been renamed *ap*^{s}.) Note that in I the number of antennal segments is 13 instead of 11. (J,K) Right and left antennae of Fused antennal segments-5 (*Fas-5*). J1–3 and K1–3 are the tarsi of the left legs of the same beetles. Note that there are various fusions and/or the distal tarsomere may be swollen. (L) Right and left antennae of fused antennal segments-3 (*fas-3*). L1–3 are the tarsi of the right and left legs of the same beetle. Note reduction in number through fusion of adjacent tarsomeres.

TABLE 4
Distribution of Fused Antennameres in the
Two Sexes of the *fas-2* Mutant in *Tribolium castaneum*

Antenna		Number	
Left	Right	Males	Females
0	0	0	2
0	3–4	1	0
4–5, 6–7	4–5	1	0
5–6	0	1	0
5–6	4–5	1	1
4–5	6–7	1	0
5–6	5–6	12	27
5–6	3–4	1	0
5–6	2–4–5, 6–7	1	0
5–6, 9–10	5–6, 9–10	1	1
4–5, 6–7, 8–9	0	1	0
5–6, 9–10	5–6	3	2
3–5	5–6	0	1
4–5, 6–7	5–6	0	1
5–6	5–6, 9–10	0	1
5–7	5–7	0	1
4–5	5–6	0	1
4–5, 6–7	4–5, 6–7	0	1
4–5	4–5	0	2
5–6, 9–10	5–6, 9–10	0	1
4–5, 6–7	4–5, 6–7–8	0	1
		24	43

Although *fas-2* is known to be in this linkage group, further work is necessary to determine its position since the data of crosses involving *fas-2,Be,* and *s* gave inconsistent results. (See data in Chapter IV, A, 4.)

d. LINKAGE GROUP V

(1) Jet (*j*, Park, 1954) is a recessive body color gene found in Dr. Thomas Park's laboratory at the University of Chicago and kindly supplied by him to the writer for further tests. It is somewhat darker than sooty (*s*) and lighter than black (*b*), differing from *b* in that it has a more reddish tinge and the appendages are less opaque. It is a good marker for this linkage group since there is no apparent reduction in viability, and penetrance is complete. The homozygous mutant can be

identified in large larvae and late pupae by the same criteria described for *b* (*q.v.*).

Eddleman (1961) found a mutation which proved to be allelic to *j* and it has been designated *j^E*. The *j^E* strain has a reduction in viability of 30% (Eddleman, 1961).

(2) Split elytra (*spl*, Lasley, 1960) is an autosomal recessive. The elytral tips are separated at their distal ends, uncovering a portion of the unsclerotized abdomen (Plate 9,M). In some genic backgrounds both the elytra and the legs may be shortened, the effect being primarily confined to the femurs, which become somewhat shorter and broader. Phenotypic expression is variable, overlapping wild type so that only about 80% of genetically *spl* beetles exhibit a recognizable separation of the elytra. Pupae of *spl* can be recognized by the fact that the elytra and membranous wings tend to curve away from the abdomen. Mortality is high among eclosing imagoes and productive life is markedly reduced in the female. The gene has been located 6–17 units away from *j* (Lasley, 1960a).

(3) Microcephalic (*mc*, Sokoloff and Lasley, 1961). A recessive with complete penetrance, variable expression, and a reduction in viability of at most 10% in various genetic crosses. The primary effect appears to be a reduction of the cranial portion of the head. Since neither the genae nor the anterior parts of the head are affected, the head appears roughly like a barbed harpoon, the genae representing the barbs. The eye is variably reduced in size: the dorsal facets seem to be absent more frequently, the remaining facets being confined to the ventral and latero-ventral portions of the head. In some cases the eye on one side may consist of a few ommatidia or they may be completely wanting. (Plate 3, C 1–3). If these beetles are selected they produce offspring with both eyes missing. It is noteworthy that if an eye color gene such as *p* is present, it becomes evident that if any part of the eye develops, an ocular diaphragm (complete with ocular foramen) of corresponding size develops under the ommatidia. Cursory examination of the larvae leads to the conclusion that the mutant can be identified in this stage only if one or both ocelli are missing. The gene is located about 25 units from *j* and about 41 units from *spl* (Sokoloff, 1962a).

Two other abnormalities have been observed in selected stocks of *mc*. The dorsal lobe of the normal compound eye has about five or six diagonal rows of ommatidia. In some *mc* beetles the dorsal component of the compound eye has a bulging appearance and the diagonal rows appear to increase to seven or eight in number. The other abnormality may be symbolized mc^{eg}. This is characterized by a growth apparently emerging from the eye (but it may really be an outgrowth of the gena which emarginates the eye producing the dorsal and ventral lobes of the compound eye). When first discovered the mc^{eg} growths were small pear-shaped structures, appearing as if one of the ommatidia had increased in length, rising above the level of the surrounding single eyes. Since then a large number of much longer, segmented growths resembling the antenna has been observed on one of the eyes of these mc^{eg} beetles (Sokoloff, 1964c).

(4) Maroon (*m*, Eddleman, 1962) is a recessive eye color mutation. The eye becomes maroon (8 pi) in pupae; in young adults the color is of the same hue, but in older imagoes the intensity develops rapidly. At the end of 7–10 days the eye appears black. Upon dissection the eye appears deep maroon ($7\frac{1}{2}$ pl). The mutant can be identified from wild type in aged adults by searching for the ocular diaphragm, a part of which remains visible in these dark-eyed beetles. This gene has been located 22 units away from j^{E} (Eddleman and Bell, 1963) and 40 units away from *mc* (Eddleman, 1964, personal communication), but its position in relation to *spl* remains to be determined.

(5) Fused antennal segment-3 (*fas-3*, Sokoloff and Ho, 1961). A recessive of good viability and penetrance (if both antennae are examined), but variable expression, found in a sample collected in a feed bin at the Zoology Department, University of California, Davis, California. One of the original (nonvirgin) females had funicular fusions and barely discernible divisions of the club segments (Plate 6,L). The F_1 were normal, and the F_2 had the appearance shown in Table 5.

Although there is a certain overlap in phenotype when compared with *fas-2* (see above) the modal number is different; *fas-3* also differs from the average appearance of *fas-1* (see below). The *fas-3* gene is located about 40 units from *j* (Sokoloff, 1965b), but its location in respect to other genes has not been determined.

(6) Fused antennal segments-3a (*fas-3a*). This appears to be a more strongly expressed allele of *fas-3*. When first found the *fas-3* mutant had only slight abnormalities of the funicle and a diffuse fusion of the club segments (see Table 5). The stock established from the original mutants has retained the original range of expression and the tarsi do not exhibit any fusions. The expression of the new allele as can be seen from Table 6 has segments 3–5, and 6–7 of the funicle fused, and in addition segment 8 of the funicle may be fused with all of the club segments, or may remain intact, separating fused blocks from the funicle and the club,

TABLE 5
Distribution of Fused Antennameres in the *fas-3* Males and Females in
Tribolium castaneum

Antenna		Number	
Right	Left	Males	Females
4–5	0	5	5
4–5	4–5	2	4
0	4–5	9	6
6–7	0	2	6
6–7	6–7	5	4
0	6–7	7	4
4–5	6–7	5	0
6–7	4–5	1	1
0	4–5, 6–7	1	0
4–5, 6–7	0	1	1
3–4	0	0	1
4–5, 9–10	0	0	1
4–5, 6–7	4–5, 6–7	3	4
4–5, 6–7	4–5	3	0
6–7	4–5, 6–7	1	0
4–5, 6–7	6–7	1	0
4–5	4–5, 6–7	1	1
		47	38

the latter consisting of fusions of segments 9–10, 10–11, or 8–11 (see, for example, Plate 6,L). The *fas-3a* allele overlaps wild type in dominance (Sokoloff, 1965c).

e. LINKAGE GROUP VI

(1) Microphthalmic (*Mo*, Sokoloff, 1960; *Mo-1*, Sokoloff, 1962, kindly supplied by David Mertz, University of Chicago). The original mutant was isolated as a nonvirgin female whose compound eyes were not fully formed: the ventral ommatidia were present, but the lateral and dorsal components of both compound eyes were missing. The F_1 were classifiable into two groups of about equal numbers of beetles with normal and abnormal eyes. The latter group exhibited a wide spectrum of abnormalities: in some the ommatidia of the laterodorsal portion of the eye were not arranged in diagonal rows as is characteristic of the arrangement in the majority of normal beetles; instead, some of the rows were incomplete, confluent with adjacent rows, or completely disrupted; in

TABLE 6

Fusions Found in the Antennae of *fas-3a* in *Tribolium castaneum*

Num-ber	Males		Females	
	Right	Left	Right	Left
1	3–5, 6–7	3–5, 6–7	3–5, 6–7	3–5, 6–7
2	3–5, 6–7, 10–11	3–5, 6–7, 8–9	3–5, 6–7	3–5, 6–7
3	3–5, 6–7, 9–11	3–5, 6–7, 8–11	3–5, 6–7, 8–9	3–5, 6–7, 8–11
4	3–5, 6–7, 10–11	3–5, 6–7, 10–11	3–5, 6–7, 8–9	3–5, 6–7
5	3–5, 6–7, 8–9	3–5, 6–7, 8–9	3–5, 6–7, 8–11	3–5, 6–7, 8–11
6	4–5, 6–7, 8–9	4–5, 6–7	3–5, 6–7, 8–11	3–5, 6–7, 8–11
7	4–5, 6–7	4–5, 6–7	3–5, 6–7	3–5, 6–7
8	3–5, 6–7, 9–11	3–7, 8–11	3–5, 6–7	3–5, 6–7
9	3–4, 6–7, 9–11	3–5, 6–7, 8–11	3–5, 6–7, 8–9, 10–11	3–5, 6–7, 8–11
10	3–5, 6–7, 8–11	3–5, 6–7, 8–11	3–5, 6–7	3–5, 6–7
11	3–5	0	3–5, 6–7, 8–11	3–5, 6–7, 8–11
12	—	3–5, 6–7, 8–11	3–5, 6–7, 8–9	3–5, 6–7, 8–9
13	3–5, 6–7	3–5, 6–7	3–5, 6–7, 8–11	3–7, 8–11
14	3–5	3–5	3–5, 6–7, 8–9, 10–11	3–5, 6–7, 8–9
15	3–5, 6–7, 8–11	3–5, 6–7, 8–11	3–7, 8–11	3–7, 9–11
16	3–5, 6–7, 8–11	3–5, 6–7, 8–11	3–5, 6–7, 8–9, 10–11	3–5, 6–7
17	3–5, 6–7, 8–9, 10–11	3–5, 6–7, 8–9, 10–11	3–7, 8–11	3–7, 8–11
18	3–5, 6–7, 8–11	3–7, 8–11	3–5, 6–7, 8–9	3–5, 6–7
19	3–5, 6–7, 9–10	3–5, 6–7, 8–9, 10–11	3–5, 6–7, 8–11	3–5, 6–7, 8–11
20	3–5, 6–7, 8–11	3–5, 6–7, 8–11	3–5, 6–7	3–5, 6–7, 8–9

others, the ommatidia appeared larger than normal and fewer in number; in still others, the dorsal and lateral portions of the eye were missing, and in a few the eyes were reduced to a small number of ommatidia—not necessarily symmetrically, and not necessarily confined to the ventral portion of the head. Rarely, one of the compound eyes was split into two widely separated components, one ventral and one lateral, consisting of a few ommatidia. In what is regarded as the extreme expression of the gene, the cranium behind the genae was reduced in size, this portion of the head appearing microcephalic; and in some beetles the head was retracted into the prothorax up to the level of the eyes. Beetles of the latter type do not survive and it is not possible to state whether they are $Mo/+$ or Mo/Mo. Beetles homozygous for pearl (p) enable one to examine the inner structure of the eye. Mo-pp beetles revealed that the ocular diaphragm, a black endoskeletal structure underlying the compound eye, is reduced in Mo beetles, the reduction of this structure paralleling the reduction of the eye. The mutant can be identified in the larva or pupa only in those individuals showing a reduction in size of the cranium. Three views of normal and *Mo* adults are shown in Plate 3, A, 1–3, B, 1–3 respectively.

In selected stocks the expression of *Mo* becomes more uniform and *mc*-like. On outcrossing, however, variability in expression of *Mo* is restored to the point that some *Mo* beetles are difficult to distinguish from wild type except after considerable familiarity with the mutant. Genetically, *Mo* behaves like a dominant with recessive lethal effects, the lethal effect probably occurring in the egg stage. *Mo* is the only gene identified so far with this linkage group (Sokoloff, 1962a).

f. Linkage Group vii

Eleven mutants have been identified with this linkage group, but eight of them are at the same locus. Linkage studies have revealed that crossing-over (which for genes associated in other chromosomes is the same for the two sexes) is unequal. This point will be treated at more length in a later section (Chapter VI, Section C,1). The relative distances given here are based on values obtained from two- and three-point crosses where the females are heterozygous for the various genes.

(1) Chestnut (*c*, Eddleman, 1961; *c⁸*, Sokoloff, 1961). These two alleles of *c* were found in unrelated material at about the same time. The *c* gene is a recessive modifying the color of the compound eye and detectable in all stages of the life cycle, except when the pupa is freshly formed and deposition of pigment in the ommatidia has not begun. The eye appears reddish-brown in young beetles but darkens to a deeper red in about 10 days. The phenotype darkens further toward the wild-type black in old beetles, but is readily identifiable. Eddleman and Bell (1963) and Sokoloff (1963a,d) independently determined that *c* was linked with *Sa* (see below) and about 40 units from it. According to Eddleman and Bell (1963) the ommatidia in this mutant have pink granules in low concentration, while the normal has blackberry-red granules in extremely heavy concentration in histological sections.

(2) Blistered elytra (*ble*, Sokoloff, 1961) is a recessive with variable expression and incomplete penetrance. It may be detected in a few pupae, but usually the blister develops at metamorphosis. The blister develops anywhere along the middle of the elytra or on the anterior (lateral) margins. In its extreme expression the blister may be as large as or larger than half the total surface of the elytron (Plate 9, O,Q). Fluid in the blisters fails to return to the body cavity before the exoskele-

ton is sclerotized, with the result that, even in old beetles, a pocket of fluid is present. Even in one individual the character may not be symmetrical, and one elytron may appear blisterless. In stocks started with beetles known to be *ble/ble* (and virgin) a large number of individuals with normal elytra may be observed (Sokoloff, 1961b). The *ble* gene is located between *c* and *Sa*, about 42–45 units from *c* and about 4 units from *Sa* (Sokoloff, 1964b).

(3) Fused tarsi and antennae (*Fta*, Sokoloff and St. Hilaire, 1962) is a dominant, with recessive lethal effects. In mildly expressed beetles the antennal segments are reduced to a total of seven with no recognizable club (Plate 5,I). Strong expression of *Fta* results in beetles with antennae consisting of, at most, two segments. The tarsi of all legs, which normally fit the formula 5-5-4 segments characteristic for the family, are reduced to at most two tarsomeres and more often to only one (Plate 5,J), and this claw-bearing tarsomere may be partly or fully fused to the tibia. In addition, the elytra in most *Fta* diverge, the divergence usually starting at the scutellum, exposing part of the unsclerotized dorsal surface of the abdomen (Plate 5,H).

Interaction of this gene with *Sa* produces a dominant synthetic lethal (Sokoloff, 1964a). *Fta* is located between *ble* and *Sa*, about 5 units from *ble* (Sokoloff, 1964b). Although viability of *Fta/+* appears to be good in linkage studies, under population conditions it declines very rapidly and requires constant selection.

(4) The *Sa* locus is represented, at the present time, by eight alleles: five are dominant with recessive lethal effects, and three are semidominant or overlap in dominance with wild type. The dominants are the following:

(*a*) Short antenna (*Sa*, Shideler, 1960), found in an irradiated population.

(*b*) Short antenna-1 (*Sa-1*, designated first as Gnarled, *Gn*, by Bunch, 1960). Found in an irradiated population.

(*c*) Short antenna-2 (*Sa-2*, originally named Distorted, *Ds*, by Dawson, 1962). Spontaneous.

(*d*) Short antenna-3 (*Sa-3*, originally called Curved Appendages, *Cua*, Sokoloff, 1961). Spontaneous.

(*e*) Short antenna-4 (*Sa-4*, Sokoloff, 1963). Spontaneous.

The semidominant or partially dominant alleles are the following:

(*f*) Short antenna (*sa*, originally called curved appendages, *ca*, Sokoloff, 1960).

(g) Short antenna-1 (*sa-1*, originally designated IW in our laboratory to avoid confusion with *ca*, Sokoloff, 1961).

(h) Short antenna-2 (*sa-2*, Dawson, 1963).

The effect of *Sa* is primarily confined to the antennae, these appendages becoming shorter than those in the normal beetle owing to a reduction in number and/or fusion of the six central segments. The number of affected segments may vary, but penetrance is complete and the expression does not overlap wild type. Some of the *Sa*/+ individuals have one or more deformed legs, but this manifestation is not reliable for classification purposes. Krause *et al.* (1962) determined that mortality of *Sa*/*Sa* occurs in the egg stage. *Sa*/+ has a reduced viability of about 15%. *Sa* takes about 2 days longer than wild type to develop to the pupa stage (Krause, 1963a).

Sa-1 and *Sa-2* have a stronger manifestation in antennal fusions and the effect on the legs is more pronounced (Plate 5,B). The data for *Sa-2* drawn from Sokoloff *et al.* (1963), may serve to illustrate this. Three strongly expressed females whose antennae had antennal segments 2–8 fused and curved, and which exhibited deformities in at least three legs, were mated to wild-type males. The F_1 were scored for leg and antennal abnormalities and sex. Since the distribution of phenotypes among male and female progeny was the same; and since it did not seem to be influenced by whatever genetic differences may have existed among mothers; and since mirror-image counterparts of beetles having the antennae asymmetrically deformed could for the most part be obtained, the following table groups the various abnormalities irrespective of sex or left- and right-handedness.

Antennal Abnormalities in *Sa-2* Individuals

Affected segments	Number of individuals
2–8, 2–8	44
2–8, 2–7	3
2–8, 2–6	1
2–8, 2–5	4
2–7, 2–7	3
2–7, 2–6	3
2–6, 2–5	1
7–8, 7–8	1
Total	60

Of the 60 beetles examined, 20 had normal legs, and the remainder had 1–5 abnormal legs. Of the total legs observed, 26 were forelegs, 28 middle, and 32 hind legs.

Sa-3 and *Sa-4* have as variable an expression as *Sa-1* and *Sa-2*. In practice, it is more reliable to examine for antennal fusions than for leg abnormalities in the identification of the mutant.

The effects of the *sa* allele are similar to *Sa* in that in *sa/sa* the antennal segments are variously fused, resulting in a curved funicular portion. It differs from *Sa* in that the femurs of all the legs become short and thick (Plate 5, C,D). If the tibiae are affected, they become variously curved. In +/*sa* the legs are normal, but the antennae of a large proportion of beetles may exhibit slight fusions (sometimes only on one antenna): the fusions occur chiefly between segments 3–4, but fusions of other segments may take place. For practical purposes *sa* and *sa-1* can be considered as incompletely recessive, but *sa-2* is a semidominant since all heterozygotes can be identified by slight antennal fusions (Dawson and Sokoloff, 1964).

g. LINKAGE GROUP VIII

(1) Antennapedia (*ap*, Englert, 1962; *ap^D* Dawson, 1962). It is worthy of note that these allelic recessive homeotic mutants were found independently (in two different laboratories), at about the same time, in two alleles of *Sa*: *ap* was found in a stock of *Sa* and *ap^D* in the (unrelated) stock of *Sa-2*. The *ap* mutant was said to exhibit "in addition to the reduction and fusion of the antennal segments, the presence of tarsal segments and the two tarsal claws on the antenna to give a leglike appearance. In its less extreme expression only partial fusion of the ninth and tenth distal segments of the antenna is evident, but the mutant is easily distinguished by the tarsal claws arising from the eleventh antennal segment" (Englert and Bell, 1963b; see Plate 7,I).

In the typical *ap^D* allele, the two basal segments are always present. The remaining nine segments of the antenna are usually replaced by what appears to be a large irregular block of sclerotized material and a usually distinct tarsus. The large block presumably represents the first four leg segments. The "antennal tarsus," when well expressed, consists of five segments plus the typical tarsal claws at the end; thus, it is similar to the tarsus of the first two pairs of legs. In addition, the spurs normally found at the end of the tibia are usually present on the mutant antenna (Plate 7,H).

The two alleles of *ap* affect the appearance of other characters: in the adult the metathorax is reduced in length to about two-thirds of the normal, and its medial surface is protuberant, giving the beetle a "humped" ventral appearance. The dorsal surface of the elytra appears more convex. In addition, the two distal tarsomeres in all three pairs of legs fail to separate completely.

The effect of *ap* or *ap^D* is recognizable as soon as the larva emerges from the egg. In normal larvae the antenna consists of four segments, the small distal one bearing a single bristle (Plate 7,J). In *ap* the distal three segments may be fused and terminate in a claw, with the result that the antenna has the appearance of a short leg (Plate 7,K).

The two *ap* alleles differ in viability: *ap* exhibits no reduction in viability from that of wild type (Englert and Bell, 1963a,b; Englert *et al.*, 1963) while the viability of *ap^D* is reduced by about 28% (Sokoloff and Dawson, 1963a).

A third allele of *ap*, *ap^s*, has recently been recovered. It is described below under the name "fused antennal segments-6" (*fas-6*). (See Section A,3,*h* in this chapter for further details.)

Englert (1963) reported lack of linkage of *ap* with any of the markers used to identify linkage groups II–VII; thus, *ap* can serve to identify linkage group VIII.

(2) Squint (*sq*, Bywaters, 1960) is a recessive gene, having semilethal effects, which prevents the formation of the ommatidia in the adult (Bywaters, 1960). The ocular diaphragm forms, and is visible through the exoskeleton, but the ocular foramen is considerably reduced (Plate 3,H). The mutant can be identified in the larva by the absence of ocelli, and in pupae if the age of the pupa is approximately known: as pointed out by Ho (1961), in 1-day-old pupae pigmented clusters of cells are already formed and visible. Absence of these cells identifies *sq* pupae more than a day old. In late pupae *sq* is identified by the presence of a pigmented ocular diaphragm visible through the pupal skin. Englert *et al.* (1963), Englert and Bell (1963b), and Sokoloff and Dawson (1963b) independently established that *ap* and *sq* are linked and about 7 units apart. Englert and Bell (1963b) have shown that in *sq* development is delayed by at least 1 day to the pupa stage and that *sq* females are not as fecund as normal females. It appears that *sq*, like *r* and *p*, is unstable: in +/*sq* heterozygotes it is possible to obtain individuals which have undergone a somatic mutation resulting in beetles with one eye normal and the other squint (Sokoloff, 1963d).

(3) Short elytra (*sh^H and D*, Ho and Dawson, 1962; *sh^s*, Sokoloff, 1962). The first of these recessive alleles was found in a "natural" population

A-1

A-2

B

C

D

E

F

G

H

I

J

K

L

M

N

O

P

collected at the University poultry farm near Berkeley, California (Ho and Dawson, 1962); the second was derived from a substrain obtained 7 years earlier from the wild-type strain designated as Chicago wild type (Sokoloff, 1962d). Tests of allelism confirmed the allelic nature of the two mutants. Both alleles are characterized by variable expression and incomplete penetrance. In sh^H ^{and} ^{D} the elytra are shorter than in sh^S: in the latter, at most, only the distal abdominal tergite is exposed (Plate 5,K); in sh^H ^{and} ^{D} up to one-third of the posterior unsclerotized surface of the abdomen may be exposed. In both alleles the elytral tips may diverge to a varying degree. Viability of sh^H ^{and} ^{D} is low, but that of sh^S is good. The latter allele was used in linkage tests and its position was established 27.8 ± 1.2 units from sq (away from ap) (Sokoloff and Dawson, 1963a).

(4) Elbowed antenna (*elb*, Dawson and Ho, 1962) is a recessive of poor penetrance and variable expression found in the same population as sh^H ^{and} ^{D}. One or both antennae may be affected: the bend of the "elbow" may be moderate to extreme, in which case the antenna forms a 45° angle. Any of the six funicular segments may act at the point of retroflexion. (Plate 7,G). Viability of extreme *elb* individuals is low, apparently as a result of the failure of these imagoes to shed the pupal skin (Dawson and Ho, 1962). The recessive *elb* was located 23 units from *sh* and subsequently lost, so that its exact position in reference to *ap* and *sq* could not be determined (Sokoloff and Dawson, 1963a).

h. LINKAGE GROUP IX

(1) Prothoraxless (*ptl*, Lasley and Sokoloff, 1960; ptl^D, Dawson, 1964), is a semidominant of variable expression overlapping wild type, and

---

PLATE 7. Various mutants in *Tribolium casteneum*. (A) Alate prothorax (*apt*): (1) pupa, (2) adult. (B) Two specimens of looped median groove (*lmg*). (C) The mutant "scar" (*sc*). Stippled area represents the deformity, a shallow depression in the metathorax. (D) Incomplete metathoracic projections (*imp*). Note the rounded posterior end of the metathorax and the sinuous median groove. (E) Megalothorax (*mgt*). (F) Jagged antecoxal piece (*jac*). (G) Two specimens of elbowed antenna (*elb*). (H) The expression of antennapedia (ap^D) in the adult. (I) The expression of antennapedia(*ap*) in the adult, after Englert and Bell (1963b). (J,K) Antennae in the normal and the ap^D larvae. (L) The phenotype of bent femur (*bf*). (M) The phenotype in bent tibia (*btt*). (N) A fairly strongly expressed "sternites incomplete" (*sti*) adult. Stippling represents unsclerotized area. (O) Two examples of "partially pointed abdominal sternites" (*ppas*). (P) The aedeagus in "emasculated" (*em*). (G,H, by permission of the Editor, *Canadian Journal of Genetics and Cytology;* I, by permission of the authors and Editor, *Annals of the Entomological Society of American.*)

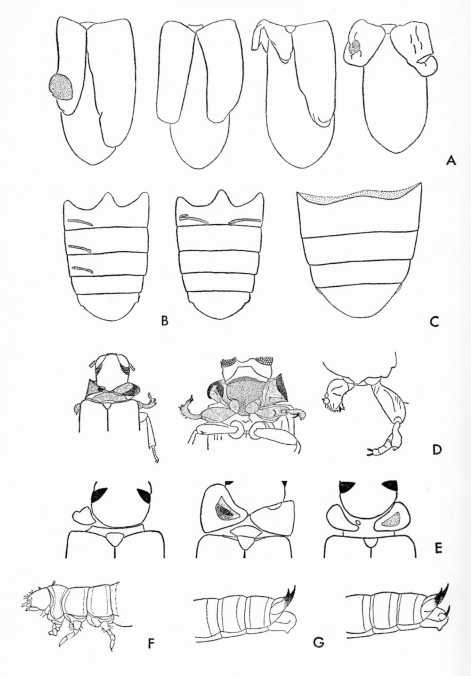

shows incomplete penetrance. It is lethal in a homozygous state, and identifiable in the earliest larval instars (Plate 8,F). In the heterozygote the prothoracic segment in the larva and the prothorax in the pupa and adult, and/or the prothoracic legs in all three stages may be affected. If only the prothorax is affected, the pronotum may exhibit a deep groove at a right angle to the midline, or it may be variously indented as if a large piece had been bitten off. In the extreme cases the pronotum is almost completely absent, and only half of the prosternum remains. If the prothorax is not affected, either leg, both legs, or no legs may be affected. If they are strongly affected, the tibia and the femur are considerably shorter and thicker, and at times the leg appears paralyzed. If legs are only mildly affected the femur becomes short and thick (Plate 8, D,E).

The homozygous mutant is more severely affected: the pronotum is completely gone, leaving the "neck" exposed. Only a small portion of the prosternum remains, and to this is attached a pair of vestigial legs in all instars. Most of the homozygous mutant larvae pupate, but many pupae die before eclosion. The eclosed *ptl/ptl* imagoes walk poorly on four legs; they cannot copulate and remain on the surface of the flour, attesting to the importance of the first pair of legs in tunneling (Lasley and Sokoloff, 1960). The *ptl/ptl* stock died out after 3 months. This mutation has been maintained by selecting mildly deformed heterozygous individuals.

Studies have failed to establish linkage between *ptl* and other markers, hence it is assigned tentatively to linkage group IX (Sokoloff, 1965b). However, the poor penetrance of *ptl* on outcrossing strongly suggests that other genes should be used for the identification of this linkage group.

i. Linkage Group x

(1) Abbreviated appendages (*aa*, Sokoloff, 1962). A spontaneous recessive of good penetrance in regard to the elytra, but variable penetrance in regard to the legs. The elytra are only three-quarters as long as in normal beetles, often divergent, sometimes starting at the scutellum,

---

PLATE 8. Various mutants in *Tribolium castaneum*. (A) Four examples of vestigial elytra (*vge*). (B) Two examples of "creased abdominal sternites" (*cas*). (C) "Missing abdominal sternites" (*mas*). (D) Dorsal and ventral views, and detail of the forelegs in prothoraxless (*ptl*). (E) Three more examples of *ptl*. (F) The phenotype of *ptl* in the larva. (G) Posterior ends of normal and "extra urogomphi" (*eu*) larvae.

but more frequently only the distal parts diverge exposing the posterior dorsal surface of the abdomen (Plate 9,N). In some *aa* beetles the legs may also be affected, the podomeres becoming shorter and thicker, but not to the extent that *ma* modifies them (Plate 9, A 1–3). Furthermore, the effect may be manifested in all the legs, in only one pair (usually the mesothoracic legs), or there may not be any perceptible change in these appendages. When the effect on the legs is strong, the antennae also appear to be shorter than normal. Upon outcrossing the effect on the legs becomes less pronounced or may even disappear, but the mutant can be recognized by the shape of the elytra. Linkage tests between *aa* and markers for chromosomes serving to identify linkage groups II–IX (*p, b, Be, mc, Mo, c, ap^D*, and *ptl*) fail to indicate linkage. Hence, *aa* is suggested as a mutant identifying linkage group X (Sokoloff, 1965b).

3. Autosomal Genes Whose Linkage Relationships Are Yet to Be Established

In this section are included a number of mutants whose linkage relationships are not known (except as noted). In part this is because they were recently found, in part because they were found in other laboratories and not yet released, and in part because of the poor penetrance of the genes. This information will be provided where available. For convenience, these mutants have been grouped according to the part of the body affected, with a category grouping those genes having pleiotropic effects on several body parts. This procedure will also be followed for *T. confusum*.

a. Genes Affecting Eye Pigmentation

(1) Ivory (*i*, Bartlett, 1961). A recessive of good penetrance and expressivity, good viability and fertility, found in an X-rayed population under selection for large body size. The eye color is said to be creamy white resembling the pearl phenotype, but is not allelic to pearl (Bartlett, 1962). Since the mutant has not been released I am not able to give a fuller description.

(2) Peach (*ph*, Dewees, 1963). Recessive, from a colony found in a food storage sack at Southern Illinois University, Carbondale, Illinois. Good penetrance and expressivity. Bicolored eye with center facets appearing reddish-pink. Not allelic with *p, c, rb, w, m*, and *i* (Dewees, 1963). This mutant has not been made available for a more complete description of its phenotype. However, Dewees (1965) now indicates it is allelic with *r* and redesignates its symbol *r^ph*.

(3) Ruby (*rb*, Dewees, 1963). Recessive found in the same population as *ph*. Good penetrance and expressivity. Produces a bicolored eye slightly lighter than chestnut. It darkens somewhat with age, but can easily be distinguished from wild type. Not allelic with *p, c, w, m, i,* or *ph* (Dewees, 1963). This mutant has not been released. Dewees (1965) states that *rb* is located in the *m* region, about 27 units from *j.*

(4) White (*w*, Eddleman and Bell, 1963). Recessive of good viability and complete penetrance. The ommatidia are said to have no pigment, resembling pearl. Histological examination reveals no eye pigment granules in *w*, while wild type has blackberry-red granules in extremely heavy concentration (Eddleman and Bell, 1963). This mutant has not been released.

b. Genes Affecting the Morphology of the Eye

Glass (*gl*, Sokoloff, 1963). Spontaneous in crosses involving *r* and M^r (for further details see Chapter VII, Section E). A recessive of good penetrance and expressivity, but lethal to semilethal in various crosses. The eye has a smooth appearance (Plate 3,G), apparently owing to the elimination only of the corneal lenses (histological sections have not been prepared). The ommatidia develop pigment depending on which eye color genes are present (*r* and normal black-eyed phenotypes have been observed). The mutant is recognizable in the pupa by a peculiar dark line across the eye, paralleling the gena(Sokoloff, 1964h).

c. Genes Affecting Body Size

Tiny (*ty*, Sokoloff and Shrode, 1962). Apart from the sex-linked *py* gene which reduces the size of the beetle without deforming it, often in single-pair matings of sibs there appear beetles strikingly reduced in size and almost as small as *py*. The number of these individual varies (Sokoloff and Shrode, 1962). When these small beetles are mated *inter se*, tiny individuals, if they appear at all, are produced only in very small numbers, the size of the remaining individuals appearing to be distributed over a wide range, indicating a very large number of factors controlling body size. Since the determination of the number of genes involved would require weighing carefully pedigreed beetles for several generations this study has not been attempted. It may be noted, however, that Bray *et al.* (1962) in their selection experiments for pupal weight obtained very good symmetrical response in increase and decrease in body weight for eight generations. The shape of the curves they obtained indicates the presence of a very large number of genes controlling body size, (possibly several hundred) but their estimates of the numbers of genes involved have not been published.

d. Genes Affecting Body Color

White appendages (*wa*, Dawson, 1961). The legs, antennae, and elytra of these morphological variants lacked pigment. Dawson (1961) believes that their inheritance is dependent on a more complex situation than that controlled by single genes. The writer has found similar variants including legs which were white but appeared nonfunctional and were probably produced by exposure of the preimaginal stages, at a critical stage of their development, to quinones released by imagoes. On the other hand, some individuals in stocks bearing body color genes exhibit antennae which appear normal except for a lack of pigment, appearing whitish. If these abnormalities in pigmentation are heritable the mode of inheritance remains obscure.

e. Genes Affecting the Prothorax

Aside from *ctp*, identified with linkage group IV, and *ptl*, with group IX, three more mutations of unusual interest affecting the prothorax have been identified.

(1) Alate prothorax (*apt*, Sokoloff and Hoy, 1964) is a case of hereditary homoeosis. It was first found by the writer in a female pupa in crosses involving *ca* and *ble*. The prothorax in this pupa was greatly enlarged and the head directed away from the body at a 45° angle. The imago emerging from it had a similarly enlarged prothorax. This proved to be but one expression of the gene. A more frequent expression was the production of elytra or membranous wing-budlike appendages which arise from the lateral edges of the prothorax about midway between the anterior and the posterior edges of this body segment (Plate 7, A–2). These growths are evident in the pupa, but they may break off upon eclosion of the imago, leaving a roughly circular, unsclerotized area. Often, in the imago, the anterior dorsal margin of the prothorax is incomplete, irregularly V-shaped, and the dorsal surface of the pronotum is variously indented. With selection, the wing-budlike processes in the pupa have been observed to form an appendage which, except for the reduction in size, resembled the elytra in shape (Plate 7, A–1). Furthermore, its wing venation closely resembled that of the elytra, and tracheae were visible. Thus, the presence of these two characters leaves little doubt that we are dealing here with a homoeotic mutant in which a pair of prothoracic wings is formed. The mutant can be recognized

in the larva only if the prothorax is enlarged (since, if prothetelous cases are excluded, wing and elytral buds develop in the pupa) or if the prothorax is irregular in shape. If so, the pigmented prothoracic tergites may become asymmetrical. Preliminary crosses suggest a semidominant mode of inheritance, but penetrance is poor and viability of "alate" individuals very low, most dying in the pupa. A mutation producing pupae with similar swollen prothoraces has recently been found in a stock of *Fta c/+c* (Sokoloff, 1965c).

(2) Megalothorax (*mgt*, Sokoloff and Hoy, 1964). Found in a stock of *Be s/+s*. Characterized by having a prothorax greatly expanded and tumorlike, usually on one side of the body (Plate 7,E). In a few cases the prothorax may exhibit a variable reduction of some of its parts resembling *ptl*. The forelegs, however, have (so far) never been observed to be affected as in *ptl*. Another difference lies in the fact that *mgt* sometimes produces a wing-budlike appendage on the affected (enlarged) side, in a manner similar to that produced by *apt*. Since viability of *mgt* is low and penetrance incomplete, tests of allelism between *mgt* and *ptl* or *apt* have not been attempted (Sokoloff, 1965c).

(3) Deflected epimera (*dep*, Hoy and Sokoloff, 1964). In the normal beetle the epimera extend medially behind the coxae of the first pair of legs and their tips lie under the sternellum (Plate 13,C). In *dep* the tip of one or both epimera may be deflected, ending somewhat posteriorly from their normal position, some showing only the posterior part of the epimeral tip behind the sternellum, others with a more pronounced deflection, lying completely behind the sternellum (Plate 13,D). In a few cases the sternellum lies *under* one or both epimeral ends. A shortening of the epimeron such as that observed in *sep* in *T. confusum* has not been observed so far (Sokoloff, 1965c).

f. Genes Affecting the Metathoracic Sternites

Aside from *ims* which has been located in linkage group IV, the following mutants have been found:

(1) Umbilicus (*umb*, Hoy and Sokoloff, 1964). A spontaneous recessive found in a stock of scar (*sc-1*), q.v., characterized by a roughly hemispherical depression on the median groove of the metathorax, about one-third of the distance between the third and second pair of legs (Plate 13,G). The depression may vary in size from barely wider than the groove to one of considerable size. It may be absent altogether in beetles known to be *umb/umb*, indicating incomplete penetrance (Sokoloff, 1965c).

(2) Dent (*dt*, Eddleman and Hudson, 1962) or dented (*d*, Ganguly, 1964) have been reported by the Purdue group as having been found in an irradiated population: *dt* was found in the F_3 generation of X-irradiated Purdue Foundation A. Eddleman's data (1962) suggested that the dent phenotype has a quantitative basis with some sex-linked gene(s). Ganguly (report by Bell, 1964) states that *d* is an autosomal recessive found in an X-irradiated population (it is not indicated whether this population was the same as that in which *dt* was found). This mutant has fair viability and uniform expressivity, with about 60% penetrance possibly being due to modifying genes. The *dt* and *d* have not been tested for allelism at Purdue, and they have not been made available to test for allelism with *umb*.

(3) Scar (*sc*, Eddleman, 1960, and *sc-1*, Sokoloff, 1963) are independent occurrences of the same or allelic mutations (Bell and Shideler, 1964). The mutant *sc* was found at Purdue and *sc-1* at Berkeley in unrelated stocks (the latter, originally called engraved metasternum, *ems*, was found in the F_3 of an X-irradiated female). They are characterized by an irregularly triangular shallow depression between the median groove of the metasternum and the pleural sclerites and anterior to the (normal) transverse groove of the metasternum (Plate 7,C). Often only one side is affected and the "scar" may be of varying size. Bell and Shideler (1964) have found that these two alleles have incomplete penetrance (70–85%) and are partially dominant to the wild-type allele. Under culturing conditions of 33°C and 70% relative humidity the direction of dominance is toward wild type, but at 35° the direction of dominance is toward *sc* (Bell *et al.*, 1964).

(4) Incomplete metathoracic projections (*imp*, Sokoloff, 1963). Spontaneous in crosses attempted to establish the position of the sex-linked M^r. A recessive of poor penetrance and variable expression characterized by the elimination or displacement of the posterior processes of the metasternum, which normally form an inverted V to accommodate the median sternal projection of the first apparent abdominal segment. In addition, the median groove is deflected to one side (Plate 7,D), and the metasternum acquires a protuberant appearance (Sokoloff, 1964h).

(5) Looped median groove (*lmg*, Hoy and Sokoloff, 1964). Spontaneous in derivatives of a sample of scar provided by A. E. Bell. The median groove in normal beetles is a shallow line extending anteriorly from the posterior end of the metasternum and terminating approximately three-fifths of the way between the third and second pair of legs. The remaining two-fifths, representing the basisternum, are not grooved. Anteriorly,

the single medial metasternal process is fused with, and covers the meso-spinasternum, a posteriorly bifurcated structure which is normally not observable because of this fusion. In weakly expressed *lmg* beetles, the median groove forms a tiny oval-shaped loop (anteroposteriorly oriented) at about the same position where *umb* forms its hemispherical depression (Plate 7,B). This loop is larger in strong expression of the character; it may be located more anteriorly and is very shiny, suggesting a delayed sclerotization of this part. Midway along the loop there may arise some secondary, short, transverse branches perpendicular to the long axis of the median groove. Furthermore, the median groove may extend to the anterior border of the metasternum, splitting the medial process into two components and uncovering the two prongs of the mesospi-nasternum (Sokoloff, 1965c).

(6) Jagged antecoxal piece (*jac*, Hoy and Sokoloff, 1964). Derived from a population traceable to an irradiated (6000 kr) male mated with an unirradiated female (but more than three generations had elapsed before it was detected). In normal beetles the antecoxal piece is a band-like area delineated by the anterior transverse grooves (which extend from the median groove laterally almost to the junction between the metasternum and the pleural sclerite), and the rear metasternal margins which assume a smooth curvilinear appearance. In *jac* either the portion including only the transverse groove, or this portion *and* the antecoxal piece are drawn anteriorly, forming a triangle with the anterior edge of the hind coxae (Plate 7,F). This mutation is not the same as scar (*sc*) since, in the latter, part of the exoskeleton *anterior* to the transverse groove is scarlike and triangular in shape (compare with Plate 7,C). Some *jac* beetles exhibit a hole between the antecoxal piece and the transverse groove. Preliminary crosses suggest a recessive mode of inheritance (Sokoloff, 1965c).

g. Genes Affecting Abdominal Structures

(1) Creased abdominal sternites (*cas*, Sokoloff, 1963). A recessive of good viability, variable expression, and incomplete penetrance, character-ized by the presence of a diagonal groove on the abdominal sternites. In extreme expression all segments possess a groove, although that in posterior sternites is considerably shorter. More often only one side of the first apparent abdominal segment bears such a groove (Plate 8,B). It fails to link with genes serving as markers for IV and VIII (Sokoloff, 1964h). By analogy with results for *T. confusum* (Sokoloff, 1964f) and *L. oryzae* (Sokoloff, 1964g), *cas* is probably linked with pearl.

(2) Elongated juvenile urogomphi (*eju*, Sokoloff, 1962). Found in the Berkeley synthetic strain marked with *s*. A recessive of variable expression: the "urogomphi" appear more blunt and fleshy rather than sclerotized. The location of these paedomorphic structures is one on each side of the anal opening. Not allelic with *ju* or *rju*, although in the same linkage group (IV) as *ju* (Sokoloff, 1965b).

(3) Reduced juvenile urogomphi (*rju*, Sokoloff, 1963). Spontaneous occurrence during attempt to establish mode of inheritance of *sta* (spikes on trochanters and antennae, q.v.). Incompletely recessive. The "urogomphi" in the adult of this paedomorphic mutation become a barely discernible pair of bristles located, one on each side, next to the anal opening. Not allelic with *ju* or *eju*, although all three genes are in the fourth linkage group (Sokoloff, 1965b).

(4) Partially pointed abdominal sternites (*ppas*, Sokoloff, 1963). Spontaneous. Discovered at about the same time in two unrelated crosses (possibility of contamination is excluded). A recessive of complete penetrance and variable expression, and viability somewhat reduced owing to inability of imagoes to shed the pupal skin. A serial homology of the abdominal sternites is attempted. So far only the second apparent abdominal segment acquires a median sternal projection (except in one individual in which the effect extended to the third apparent abdominal segment as well) mimicking the shape of the first apparent abdominal segment (Plate 7,O). As a result of the fact that the newly created medial sternal projection lies over the posterior edge of the preceding segment, mechanical difficulties are encountered by these beetles during their development so that the edges of adjacent sternites do not overlap, and a broad unsclerotized area is present between these segments. There is some variation in the size of the medial sternal projection and the extent to which the intersegmental surface fails to sclerotize (Sokoloff, 1964h). (Compare the phenotype of this mutation with that of Plate 11,D, showing the sex-linked recessive *pas* in *T. confusum*).

(5) Missing abdominal sternites (*mas*, Hoy and Sokoloff, 1964). A recessive of good penetrance, expressivity, and viability. Cursory examination of the mutant reveals only four instead of five apparent abdominal segments, with a considerable unsclerotized area in front of the apparent second abdominal segment, and only a slight displacement forward of the whole abdomen, leaving but a tiny gap between the elytral edges and the lateral margins of the abdomen. Dissection of the beetles reveals that abdominal segments I and II, normally feebly sclerotized structures located dorsal to the coxae, are completely gone. The third abdominal

segment is present but greatly reduced and only feebly sclerotized. The median anterior process of this segment is practically the only thing that identifies it from the apparent second segment at the midline. Lateral to the median process the segment forms some pocketlike recesses which resemble the recesses into which the hind coxae fit, but they are only about one-fourth as wide as the second apparent abdominal segment (Plate 8,C). The character is not recognizable in the pupa because the elytra cover this area. It is not known at this writing whether the mutant can be recognized in the larval stage (Sokoloff, 1965c).

h. Genes Affecting the Antennae

(1) Fused antennal segments-1 (*fas-1*, Sokoloff, 1960). This was the first of a series of mutants having the antennameres of the funicle and/or club fused. They were given the same name and a number in the order of their discovery without in any way implying allelism. It has already been pointed out above that *fas-2* and *fas-3* are in linkage groups IV and V, respectively. Mutant *fas-1* is not allelic to either of these genes. This gene, which appeared spontaneously in the Chicago stock, is a recessive of good viability, incomplete penetrance, and variable expression. Adjacent segments of the funicle and/or the club are fused, sometimes only on one antenna (Plate 6,B), and often asymmetrically, as Table 7 (from Sokoloff, 1962d) shows.

The modal peak for expression in the two sexes is in those individuals exhibiting fusions of segments 4–5 of the funicle and 9–10 of the club, with a secondary peak showing fusions only of segments 9–10 of the club. Linkage of *fas-1* has not been established, but it segregates independently from *j* and *c*, markers for linkage groups V and VII (Sokoloff, 1964, unpublished).

(2) Fused antennal segments-4 (*Fas-4*, Sokoloff, 1963). This mutant resembles *fas-1* and *fas-2* in having segments 4–5, and/or 5–6 and 9–10 affected (Plate 6,F). Indeed, if single individuals are compared, it would be difficult to decide whether they belong to *fas-1*, *fas-2*, or *Fas-4*, since there is much overlap, as can be seen by comparing Table 8 with Tables 4 and 7.

However, the majority of *Fas-4* beetles have segments 3–4 and 5–6 of the funicle fused in the same individual, something which rarely happens in *fas-1* and *fas-2*. Tests of allelism indicate *Fas-4*, a dominant mutation overlapping wild type, is not allelic with *fas-1*, *fas-2*, *fas-3*, *Fas-5*, or *fas-6*. It appears that penetrance of *Fas-4* is more complete than that of other *fas* mutants (Sokoloff, 1965c).

TABLE 7
Antennal Fusions Found in a Sample of *fas-1* in *Tribolium castaneum*

Antenna		Number	
Left	Right	Males	Females
0	0	1	1
4–5	0	2	1
0	4–5	0	2
4–5	4–5	1	2
5–6	0	0	1
4–5, 9–10	0	1	0
4–5, 9–10	4–5	2	3
4–5, 9–10	9–10	7	8
4–5, 9–10	4–5, 9–10	25	4
0	4–5, 9–10	0	1
4–5	4–5, 9–10	1	2
9–10	4–5, 9–10	5	2
0	9–10	0	4
9–10	0	1	0
9–10	9–10	16	6
4–5	9–10	3	2
9–10	4–5	1	2
5–6	5–6, 9–10	1	0
5–6, 9–10	9–10	2	0
4–5–6, 9–10	4–5–6, 9–10	1	0
4–5–6, 9–10	4–5, 9–10	1	0
4–5, 9–10	4–5–6, 9–10	1	0
4–5	4–5, 8–9–10	3	0
		75	41

(3) Fused antennal segments-5 (*Fas-5*, Sokoloff, 1963). This mutant resembles *fas-1* and *fas-2* and *fas-3* in that segments 4–5, 5–6, or 6–7 may be affected (Plate 6,J,K). It differs, as shown in Table 9, in that several consecutive segments (3–8 or 6–8, for example) may be fused. This gene also affects the tarsi, adjacent tarsomeres becoming fused and/or the distal one enlarged. *Fas-5* is not allelic with *fas-1, fas-2, Fas-4*, or *fas-6*. It behaves as a dominant, with some overlap with wild type in heterozygotes of some crosses (Sokoloff, 1965c).

(4) Fused anntennal segments-6 (*fas-6*, Sokoloff, 1963). This mutant was found in the course of testing linkage relationships of antennapedia (*ap^D*). It is unusual in that the number of antennameres exceeds the normal number of 11 in some individuals, as can be seen in Table 10 and Plate 6,H,I, which shows an example of antennae showing 13 seg-

ments. The antennae of these individuals are greatly modified, forming a clublike structure which results from the fusion of segments 8–11 or 9–12. In many of these individuals there is evidence of clawlike structures at the tip of the antennae which are evidently produced by the ap^D gene. This mutant differs from ap^D in that the antennae are not leglike, the metathorax is not shortened, and the distal segments of the tarsus are completely separated, which is not true for the ap^D mutant. The *fas-6* gene is not allelic with *fas-1*, *fas-2*, *fas-3*, *Fas-4*, or *Fas-5*. It is recessive to wild type, but partially dominant to, and allelic with, ap^D. Hence it has been renamed ap^s (Sokoloff, 1965c).

(5) Deformed (*Df*, Eddleman, 1961) is a dominant with recessive lethal effects affecting the last four to seven distal segments (Plate 6,D).

TABLE 8
Antennal Fusions Found in a Sample of *Fas-4* in *Tribolium castaneum*

Males		Females	
Right	Left	Right	Left
3–4, 5–6	3–4, 5–6	4–5	4–5
4–5	4–6	4–5	4–5
4–6	4–6	4–5	4–5
3–4, 5–6, 9–10	3–4, 5–6	4–5	4–5
3–4	3–4, 5–6	4–5	5–6
3–4, 9–10	3–4, 9–10	4–5	5–6
3–4, 9–10	3–4	5–6	5–6, 9–10
3–4	3–4	4–6, 9–10	5–6, 9–10
3–4, 9–10	3–4, 5–6	5–6, 9–10	5–6, 9–10
3–4, 5–6	3–4, 5–6	5–6	5–6
4–5	4–5	4–5, 9–10	5–6, 9–10
3–4, 9–10	3–4, 9–10	4–6, 9–10	4–6
3–4, 5–6, 9–10	3–4, 5–6, 9–10	5–6, 9–10	5–6
3–4, 9–10	3–4, 9–10	4–6	5–6, 9–10
3–4, 5–6	3–4, 5–6	4–6, 9–10	4–6
3–4	3–4, 5–6	4–6, 9–10	4–6
5–6, 9–10	5–6, 9–10	4–6	4–6
3–4, 5–6	3–4, 5–6	4–5, 9–10	4–6, 9–10
3–4, 5–6, 9–10	3–4, 5–6, 9–10	4–6	4–6
3–4, 9–10	4–6	4–6, 9–10	4–5
3–4, 5–6	3–4, 5–6	4–6	4–6
3–4, 9–10	3–4, 9–10	3–4, 9–10	3–4, 9–10
3–4, 5–6, 9–10	3–4, 5–6, 9–10	3–4, 5–6, 9–10	3–4, 5–6, 9–10
3–4, 5–6, 9–10	3–4, 5–6, 9–10	5–6	5–6
3–4, 5–6, 9–10	3–4, 5–6, 9–10	4–6, 9–10	4–5

TABLE 9
Antennal Fusions Found in a Sample of *Fas-5* in *Tribolium castaneum.*

Males		Females	
Right	Left	Right	Left
6–8	4–5, 6–7	4–5, 6–8	4–5, 6–8, 10–11
4–5, 6–8	4–5, 6–7	6–8	4–5, 7–8
4–5, 6–7	4–5, 7–8	4–5, 6–8	4–5, 7–8
4–5	4–5	4–5	4–5
5–6, 7–8	4–5, 7–8	3–4, 5–6, 7–8	3–4, 5–6, 7–8
4–5, 7–8	4–5, 7–8	4–5	4–5
6–8	5–8	4–5, 6–8	4–5, 7–8
4–5	4–5, 7–8	4–5	0
4–5, 7–8, 9–10	4–5, 6–8	4–6	4–5
4–5, 7–8	4–5, 7–8	4–5, 6–7	4–5, 7–8
4–5, 7–8	4–5, 6–8	4–5, 6–8	4–5, 7–8
4–5	4–5	4–5	4–5
4–5, 7–8, 10–11	4–5, 7–8	4–5	4–5
5–6	4–5	4–5, 6–8	4–5, 6–8, 9–10
3–8	3–6, 7–8	4–5, 6–8	4–5, 6–8
4–5, 6–7	4–5, 6–7	4–7, 7–8	4–5, 6–8
4–5, 6–8	4–5, 6–8	4–5	4–5
3–8	4–5, 6–8	4–5, 6–8	4–5, 6–8
4–5	4–5	3–8	4–8
0	4–5	4–5, 6–8	4–5, 6–8
4–5	4–5	4–5, 6–8	4–5, 6–8
4–5, 6–8	4–5, 6–8	4–5, 6–8	4–5, 6–8
4–5, 7–8	4–5, 6–7	4–5, 7–8	4–5
4–5, 6–7	4–5, 6–8	4–5, 6–8	4–5, 6–8
4–5, 6–7	4–5, 6–7	5–6	5–6

According to Eddleman (1961) these segments become fused into various shapes: the antenna may "have the shape of a ten-pin; be bent sharply, or be deeply incised near the tip" (Eddleman, 1961). The mutant has not been released to determine its linkage relationships.

(6) Branched antenna (*bra*, Dawson, 1962a) was found three times by Dawson in widely separated derivatives of *Sa-2* and at least three additional times by other personnel in the Berkeley laboratory. Dawson (1962a) concludes that it is probably a phenodeviant, characterized by having one to four branches on one or both antennae (Plate 10,P). The three-club segments are most frequently branched, but the branching may occur anywhere on the antenna. Viability seems not to be reduced. The frequency of expression in the F_2 of the two cases investigated was

1%. Through selection and inbreeding the frequency increased to 35%, but there was a rapid drop in fertility. Outcrosses result in a decrease in penetrance.

(7) Pectinate antenna (*pec*, Hoy and Sokoloff, 1964). Spontaneous in selection vials involving *ptl*, *ap*, and *sq*. Fusions of this incompletely recessive mutant are primarily confined to the funicle and club, although

TABLE 10
Antennal Fusions and Numbers of Segments Found in a
Sample of *fas-6* in *Tribolium castaneum**

Males		Females	
Right	Left	Right	Left
4–5, 6–11c	7–8, 9–11c	8–11c	8–11c
6–8, 9–11	9–11	4–5, 8–11c	8–11c
6–8, 9–12	4–5, 9–11	6–7, 8–11c	8–11c
4–5, 6–8, 9–12	4–5, 7–11c	4–5, 8–11c	7–8, 9–11c
4–5, 8–10c	6–7, 8–11c	7–8	0
9–10	9–11	9–10	9–10
7–8, 9–11c	4–5, 7–8, 9–11c	4–5, 8–11	7–8, 9–11
9–10	10–11	5–6, 8–11c	5–6, 8–11c
9–11c	9–11	7–8, 9–11c	7–8, 9–11c
7–8, 9–12	7–8, 9–12	7–8, 9–12c	9–11
7–8, 10–11c	7–8, 9–11	7–8, 9–11	6–8, 9–11c
7–8, 9–11	7–8, 9–11	4–5, 7–8, 9–10	4–5, 6–7, 8–11
7–8, 9–12	7–8, 9–12	0	9–10
0	7–8	7–8, 9–12	4–5, 7–8, 9–13
0	10–11	0	7–8
4–5, 7–8, 9–10	7–8, 9–11	7–8, 10–11	7–8, 9–12
7–8, 9–10, 11–12, 13	4–5	7–8, 9–11	4–5, 7–8, 9–13
5–6, 9–10	7–8	4–5, 7–8c	7–8, 9–10
9–10c	4–5, 9–10c	7–8, 9–10	7–8, 9–11
4–5, 6–11c	4–5, 7–11c	7–8, 9–11c	7–8, 9–11c
4–5, 7–8, 9–11c	4–5, 7–8, 9–11c	7–8, 9–10	8–9, 10–11c
4–5, 7–8, 9–12c	4–7, 8–10c	3–4, 9–11	3–4, 6–7, 8–10, 11–12
4–5, 8–11, 12	6–7, 8–10, 11c	7–8	0
4–5, 7–8, 9–11c	4–5, 8–11	4–5, 6–8, 9–11	4–5, 6–8, 9–11
4–5, 8–11c	4–5, 7–8, 9–10c	7–8, 10–12c	7–8, 11–12
0	9–11	4–5, 7–8, 9–11c	4–5, 9–11c
7–8, 9–11	0	9–10	7–8, 10–11
4–5, 6–8, 9–10c	4–5, 6–8, 9–10c		
6–7, 8–11c	6–7, 8–11		
4–5, 7–8, 9–10c	4–5, 7–8, 9–11c		

* The letter "c" denotes that the distal segments had clawlike bristles.

TABLE 11
Fusions Found in the Antennae of *pec* in *Tribolium castaneum*

	Males		Females	
	Right	Left	Right	Left
1	4–5	4–5	4–5	4–5
2	9–10	7–10	5–6, 7–8	4–7, 8–9
3	5–6	5–6, 7–8	4–5	3–5, 7–8
4	2–7	2–8	3–4	4–5, 6–7
5	0	4–5	4–6	4–6, 7–8
6	4–5, 6–7	4–5	3–5, 6–7	3–8
7	0	4–5	4–6	4–6
8	3–7	3–6	4–5, 7–8	4–5, 7–8
9	4–5, 6–7	5–6	3–4, 5–6, 7–8, 9–10	3–6, 7–8
10	0	9–10	3–6, 7–8, 9–10	4–5, 7–8, 9–10
11	4–5, 6–10	5–6, 8–10	4–5	4–5
12	4–5	0	2–5	2–5
13	3–4, 5–6	5–6	4–7	4–8
14	7–8	7–8	—	—
15	7–8, 10–11	9–10	—	—

rarely the scape may also be fused to other distal segments. One of the expressions of *pec* is complete fusion of several antennameres on one side, but partial segmentation on the other side of the antenna, producing a structure remotely resembling the teeth of a comb (Plate 6,G). The antennameres of the funicle and club may be greatly swollen. The distribution of fusions in the original sample of beetles is as shown in Table 11. Since this mutant was just discovered, tests of allelism between *pec* and the various mutants described under the general name of fused antennal segments (*fas-1* to *fas-6*) have not been carried out, but the phenotype of these beetles is so different that it is probably a distinct gene (Sokoloff, 1965c).

(8) Spatulate antenna (*Spa*, Sokoloff and Hoy, 1964). Spontaneous in attempts to select a *ju ct; c sa* stock. Preliminary crosses suggest it is an autosomal dominant with recessive lethal effects. The phenotype of this mutant is different from that produced by *Df*, *Fta*, and *Sa* (including its dominant, semidominant, or incompletely recessive alleles). As shown in Table 12, the ten males and ten females scored exhibit fusions in the funicle and/or the club (Plate 6,F). In addition, the fused club may resemble a small scoop or spatula. Both antennae must be examined since penetrance appears to be unequal in the two sides of the body. Badly deformed *Spa* may be identified in the pupa (Sokoloff, 1965c).

(9) Elbowed antenna-1 (*elb-1*, Hoy and Sokoloff, 1964). Spontaneous in a stock of *Ds/+*; *aa/aa*. An autosomal recessive of incomplete penetrance and fairly uniform expression. The antennae exhibit a retroflexion, with the club directed toward the head (Plate 7,G), the point of retroflexion in the limited sample so far examined being located at the seventh antennamere, which is smaller than the preceding segment. Probably allelic to the now extinct mutant discovered by Dawson and Ho (1962), and found to be linked with *sh* in a linkage group VIII (Sokoloff and Dawson, 1963a).

A recent contribution to the *Tribolium Information Bulletin* by Bell (1965) adds the following two to the growing list of antennal mutants:

(10) Bead (*bd*, Shideler.) Autosomal recessive appearing spontaneously in our Chicago Inbred Line at generation 51. Antennal segments are elongated and reduced in size giving a "beaded" appearance. Frequently, the antennae are lighter in color (blond) than the wild type. Expression is variable with about 75% penetrance.

(11) Deformed antenna (*da*, Shideler.) Autosomal recessive found in a *mc m j* marker stock. The three terminal segments are fused as in the *paddle* mutation. Variable expression with about 90% penetrance.

i. GENES AFFECTING THE ELYTRA

The elytra in normal *Tribolium* imagoes, according to El Kifl (1953), cover the main lateral regions of the mesonotum, the entire metanotum, and all the abdominal tergites except the posterior portion of the seventh one, which is the last visible tergite, and they protect the membranous wings. Their anal margins meet middorsally, forming the "dorsal

TABLE 12
Antennal Fusions in a Sample of *Spa* in *Tribolium casteneum*

Number	Males Right	Left	Females Right	Left
1	5–6, 10–11	6–8, 9–10	4–5, 6–8, 9–11	4–11
2	7–8, 9–11	9–11	7–8, 9–11	10–11
3	3–8, 9–11	5–8, 9–10	9–11	5–8, 9–11
4	10–11	6–8, 9–11	5–6, 9–11	4–8, 9–11
5	6–8, 9–11	9–11	7–8, 9–11	5–6, 7–8, 9–11
6	10–11	0	5–6, 10–11	5–6, 10–11
7	6–7, 10–11	10–11	7–11	6–9
8	6–8, 9–11	9–11	3–4, 6–8, 9–11	7–8, 10–11
9	10–11	10–11	5–8, 9–11	5–6, 9–11
10	4–8, 9–11	5–8, 9–11	0	4–8, 10–11

suture," and they do not diverge posteriorly. The anal edge of the left elytron has a narrow unsclerotized portion which forms a groove which accommodates a band arising from the unsclerotized portion of the anal margin of the right elytron. The elytra are much sclerotized, with eleven longitudinal straight and nonbranching veins. Between the veins, the cells are arranged into ten longitudinal furrows or striae. Abnormalities concerning these elytral veins have been observed in only one or two cases in *Tribolium* and several cases in *Latheticus*. In the latter, a number of beetles have been observed with the secondary mediocubital vein (*mcu*) originating in common with the media; at a point about one-eighth of the elytral length, the veins separate and *mcu* continues along its normal course between the postmedial and the antecubital striae. Fusions of other veins may occur. These types of abnormalities have proved to be developmental accidents, but it is likely that others will prove to be under genetic control. It may be pointed out that in *Tribolium* the longitudinal veins are best seen shortly after eclosion of the imagoes, even in a beetle having as light a body color as in *Latheticus*. Another interesting abnormality has been seen in heavily etherized beetles: the longitudinal veins, instead of being straight, acquire a sinuous appearance (which comes as a surprise to the writer, since he always thought that these were rigid, inelastic structures).

The membranous wings in tenebrionids are completely hidden by the elytra. Any abnormalities involving these wings are practically beyond the reach of the investigator. For example, a notched condition was found in a young beetle which had not folded its wings under the elytra. This beetle was bred and some F_1 individuals sacrificed to examine the membranous wings. They were all normal. The process was repeated in the F_2 and the wings of these were also normal. Examination of these wings, by necessity, requires the deflection of the elytra, and often these sclerotized structures break off or forceps may injure the unsclerotized tergites.

Often the membranous wings of teneral adults have huge blisters. Some of these are probably produced when the beetle attempts to free itself of the pupal skin, and the hemolymph soon thereafter finds its way into the body cavity. Some blisters are probably genetic in origin, and in these beetles the blisters hold fluid for a longer time, with the result that the elytra harden out of shape. In fact, examination of the pupae of such mutants as "akimbo" in *T. castaneum* or "thumbed" in *T. confusum* reveals that the elytra already are displaced from their normal position by a blister of varying size present in the membranous wings.

A similar explanation probably holds true for a condition called "warped elytra," but pending a more thorough study which would entail examination of pupae or the membranous wings of teneral adults, they are listed here as phenodeviants.

The elytra are sensitive to various environmental factors. Studies by Slater and collaborators at Donner Laboratory, University of California, Berkeley (see, for example, Amer *et al.*, 1962; Beck, 1962, 1963; Beck and Slater, 1961; Beck and Manney, 1962; Slater *et al.*, 1961, 1963, 1964), have shown that *T. confusum* incubated at 30°C are least affected in regard to the development of phenocopies. At this temperature the elytra were divergent in 1.6% of the cases. At 38°C the frequency was increased to 4.5%. Pupae irradiated at 1200 r and kept at 30°C produced 15.2% abnormal individuals, but the frequency of abnormals increased to about 60% if the pupae were reared for three days at 38°C, and to about 68% if pupae were allowed to complete their development at this temperature. Amer (1963a,b) found that magnetic fields have protective effects against irradiation injury: pupae were irradiated with 1200 r and one group placed between the poles of a permanent magnet with a field of 3.6 kilogauss. Another group was introduced between the poles of a dummy magnet of the same geometry. Both groups were incubated at 38°C. The second group showed 93.4 ± 1.4 individuals with abnormal elytra, while the first showed only 54.1 ± 5.3 abnormals.

Roth and Howland (1941) exposed preimaginal stages of the flour beetle to gaseous quinones derived from adult *T. confusum*. Among the abnormal beetles, some were found with divergent elytra.

Thus, the finding of beetles with abnormal elytra does not necessarily mean that the abnormality is heritable, and many such beetles tested showed that the condition had no hereditary basis. In preceding pages the sex-linked mutations *dve, ma, pok,* and *te;* the autosomal genes *spl,* identified with linkage group V; *Fta* and *ble,* identified with VII; and *sh,* with VIII, have been described. The following are either phenodeviants or abnormalities with affected elytra under the control of autosomal recessive genes, but the linkage group has not been identified.

(1) Droopy elytra (*dre,* Sokoloff and Lasley, 1960) is probably a phenodeviant since matings of *dre* × *dre* produce a large number of normal beetles, and the frequency of expressed abnormalities can be increased by selection. The elytra fail to meet over the whole length of the abdomen and present a variable degree of droopiness (Plate 9,G). Viability is greatly reduced, presumably because of dehydration (Lasley and Sokoloff, 1961).

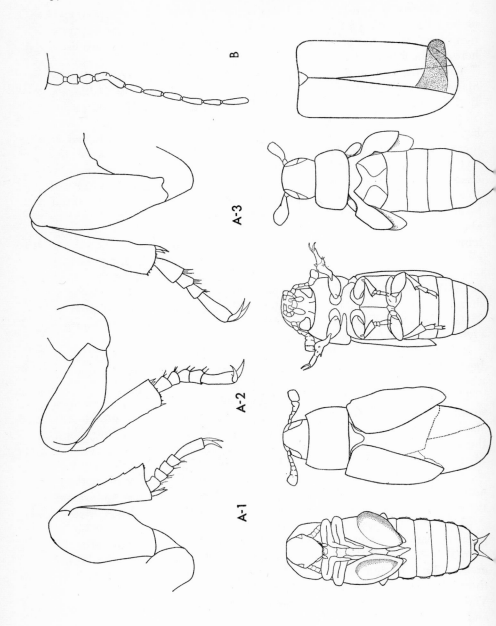

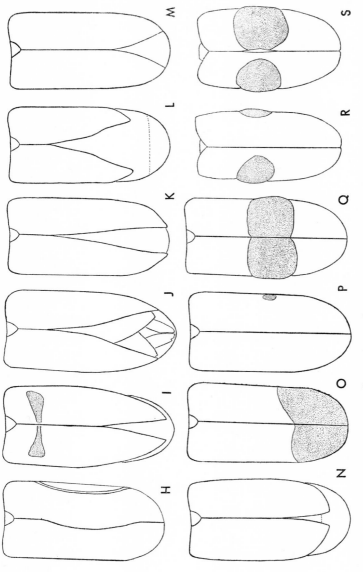

PLATE 9. For descriptive legend see p. 66.

(2) Warped elytra (*we*, Sokoloff, 1960) is probably a phenodeviant, but a further check of the membranous wings for the presence of blisters in teneral adults is necessary. The elytra meet, but both are displaced to the left or to the right of the midline, or one or both are lifted away from the abdomen producing a wavy structure (Plate 7,H). Frequency can be increased by selection, but drops on outcrossing. Viability is reduced, presumably because of loss of moisture (Sokoloff, 1960c).

(3) Vestigial elytra (*vge*, Sokoloff and Lasley). At least two independent occurrences have been observed, one in a wild-type (Chicago) strain and another in a black strain. It is lethal to semilethal in the imago, probably because beetles dehydrate very rapidly when the unsclerotized surface of the abdomen is exposed. Probably a recessive, but the stock was lost owing to the poor viability of *vge*. The elytra may be almost completely wanting (a condition which is relatively rare), or may be shortened, sometimes asymmetrically. The tips of the elytra may appear cuneiform in shape, or they may exhibit deep clefts. Sometimes the elytra curve ventrally, drooping on each side of the abdomen. A blister may develop on one or both elytra. (For range in phenotypes see Plate 8,A.) In some matings larval mortality is high, suggesting at least a semilethal effect at that stage (Sokoloff and Lasley, 1961).

(4) Pointed elytra (*pe*, Sokoloff, 1961). Spontaneous in *Mo b p* stock. A recessive of complete penetrance but variable expression: the elytra are typically divergent resembling *dve* in that the divergence often starts at the scutellum, but they differ from *dve* in that they are narrower than normal (Plate 9,I). In a fairly large proportion of *pe* the membranous wings appear affected, probably by a blister, and one or both elytra may be lifted away from the abdomen (Sokoloff, 1963e).

PLATE 9. Various mutants (chiefly in *Tribolium castaneum*, except where noted). (A1–3) The phenotype of the fore-, middle-, and hind legs in strongly expressed abbreviated appendages (*aa*). (B) An example of "crooked antenna" (*cka*) in *Cryptolestes turcicus*. (C) The phenotype of the pupa in the "rough" (*ro*) mutant. (D–E) The mutant "troll" (*tro*) viewed from the dorsal and ventral aspect. (F) Dorsal view of the "sleek" (*slk*) mutant. (G) The droopy elytra (*dre*) phenodeviant. The stippled area represents part of the membranous wing. (H) The warped elytra (*we*) phenodeviant. (I) One of the expressions of pointed elytra (*pe*). (J) An example of pokey (*pok*). (K) A mild expression of ballooned (*bal*). (L) An extreme expression of abbreviated appendages. (M) One of the expressions of split (*spl*). (N) A specimen of "concave split" or *aa-1*. (O–Q) Three examples of blistered elytra (*ble*). (R,S) Two examples of blistered elytra in *T. confusum*.

(5) Ballooned (*bal*, Dawson, 1963), is a recessive of variable expression. In the pupa the elytra and membranous wings may lie in the normal position, or they may be lifted away from the abdomen exposing the tarsi and tibiae of the hind legs. In the adult the elytra are usually widely split beginning at the scutellum and they diverge at the posterior end, completely exposing the abdomen. The elytra appear shorter by at least the equivalent of one abdominal segment and narrower than normal (Plate 9,K). The membranous wings exhibit a huge blister in teneral adults, but in older imagoes the fluid is withdrawn into the body cavity. The tips of the membranous wings remain extended and they seem to curl upward beyond the tips of the elytra. Not allelic with *pe*, *spl*, *she*, or *aa* (Dawson, 1964a).

(6) Curly elytra (*cye*, Sokoloff, 1964). Recessive. Spontaneous in *sa +/+ ble* sib crosses. In the adult *cye* the elytra are only about two-thirds as long as in the normal (resembles Plate 9,L). They exhibit variable expression: they may curve upward, or may appear splitlike or short-elytralike. In some pupae the elytra are curly, separating from the membranous wings which remain in their normal position (Sokoloff, 1964h). According to tests kindly carried out by Dr. Peter S. Dawson it is not allelic with *aa*, *she*, or *pe*.

(7) Concave split (*cspl*, Sokoloff and Lasley, 1960). Recessive. Resembles *spl* but is not allelic with this gene. In the pupa the elytra may be widely separated, and the hind tarsi extend beyond the elytral tips. In some pupae the elytra and membranous wings curve away from the body. In the adult the elytra may be split and short; or they may not exhibit any divergence, but they may not reach the end of the abdomen to cover it completely (Plate 9,N). In this sense *cspl* resembles *she* (Sokoloff, 1965c).

(8) Split, curved elytra (*spce*, Sokoloff and Rodriguez, 1963). Resembles *cye*, or *cspl*, in having the elytra and membranous wings drawn away from the tarsi in the pupa. In the adult the elytra are split and/or short (Sokoloff, 1965c).

(9) Banjo (*bj*, Lasley and Sokoloff, 1960). Recessive. Spontaneous in Chicago wild type. The elytra in the adult may be widely divergent and drooping at the sides (resembling Plate 9,L), or they may meet at the midline but be short, not reaching the tip of the abdomen (Sokoloff, 1965c).

(10) Akimbo (*akb*, Hoy and Sokoloff, 1964). Derived from a population traceable back to an irradiated (6000 kr) male mated to an unirradiated female. Several (more than three) generations later two females

and one male were found with *rt* (reduced tibia, described in the next
section). Their descendants produced a number of mutations including
sta-like, jagged antecoxal piece (*jac*), and *akb*. This mutant resembles
thu (Plates 5,E,F and 14,K,L) in *T. confusum*: the elytra are variously
raised above the abdomen, exposing the posterior abdominal segment
to a varying degree. At a point one-fifth to one-sixth of the elytral length
behind the anterolateral margin of the elytra and the scutellum there
is a notable crease or depression as if the elytra had been pushed down
at this point. The elytra in *akb* usually meet at the midline, but in a
small number of mutants they may be divergent, starting at the
scutellum. Examination of the pupa reveals that the primary cause of
this abnormality is the development of a blister in the membranous wings
which displace the elytra (Plate 5,F). Apparently the hemolymph in
these blisters is not withdrawn into the body cavity before the beetle
sclerotizes, and this produces the misshapen elytra. Preliminary crosses
suggest a recessive mode of inheritance (Sokoloff, 1965c).

(11) Rough (*ro*, Hoy and Sokoloff, 1964). This recessive mutant was
found in a vial being selected for an abnormal condition affecting the
spinasterna (which proved not to have any genetic basis). It is characterized by the appearance of sizable uni- or bilateral blisters in the
elytral buds of the pupa, which interferes with normal eclosion of a
large number of imagoes, resulting in their death. Individuals not exhibiting a blister in the pupa may develop one at eclosion (identical with
Plate 14,G), or the elytra may have a roughened appearance. The phenotype of *ro* in this species resembles in every way the *ro* mutant in
T. confusum (Sokoloff, 1965c).

(12) Bent elytral tips (*bet*, Sokoloff and Hoy, 1964). Found in a stock
of *ap bf*. A recessive of variable expression and incomplete penetrance.
In the pupa the elytra are variously separated from the membranous
wings (which remain in their normal position over the legs) and in a
few cases the tips may be folded under resembling *te* (Plate 2,F). In
the imago the elytra are variously divergent, sometimes starting at the
scutellum and the tips of one or both elytra may be bent downward
(but usually not folded under, as in *te* imagoes) and they may bear
a blister of varying size (Sokoloff, 1965c).

(13) Elongated elytra (*ele*, Sokoloff and Hoy, 1964). This autosomal
recessive mutant was found in two unrelated stocks. It resembles in every
way the *ele* mutant in *T. confusum* (Plate 14,O). Females with extremely
long elytra (extending beyond the posterior end of the abdomen by the
equivalent of the length of the posterior segment) often remain sterile,

probably due to the fact that the elytra are too long to allow successful copulation.

Tests of allelism between the various elytral mutants have largely been completed only recently. Only one mutant has proved to be allelic to others reported thus far: *cspl* is allelic with *aa*. Hence, *cspl* is changed in designation to *aa-1*. The *aa-1* allele differs from *aa* in the fact that the legs are not as strongly affected, even in the stock, as in the *aa* mutant.

j. GENES AFFECTING THE LEGS

Aside from the pegleg (*pg*) and the deformed legs (*dfl*) mutants identified with linkage groups II and IV, the following genes have been found to affect only the legs:

(1) Bowleg (*bl*, Reynolds, 1964). A recessive of variable expression and 70–75% penetrance, apparently susceptible to the influence of modifiers. Primarily the hind legs are deformed with tibiae twisted or bowed, but sometimes the effect can be observed in the other two pairs of legs. Viability is fair to good (Reynolds, 1964). This mutant has not been released by the Purdue group so it is not possible to compare it with other mutants having legs affected.

(2) Bent tibia (*bt*, Dawson, 1961). Possibly a phenodeviant since it is present at low frequency in the CS synthetic strain. Matings of *bt* × wild type produce a small proportion of *bt* progeny; hence it overlaps wild type. Mass matings of *bt* × *bt* give *bt*: + ratios of 1:1 to 2:1. Many *bt* beetles fail to shed the pupal skin from the deformed legs. The bend occurs just below the tibiofemoral joint. The femur may be curved and/or shortened. The tarsi are not affected, but the antennae may exhibit an "elbowed" condition. The legs are often broken off at the point of bend. Viability is reduced (Dawson, 1964a).

(3) Bent tibia (*btt*, Sokoloff, 1963). Recessive of poor penetrance and variable expression. In extreme expression the tibia of the hind legs has a double bend near the tibiofemoral joint, resulting in an apparently shorter leg (Plate 7,M). Detectable in the pupa (Sokoloff, 1964h).

(4) Bowed femur (*bf*, Sokoloff, 1963). Spontaneous. A recessive of variable expression: the femur on the third pair of legs is curved to a variable degree, the curvature paralleling that of the abdomen when the legs are resting on this part of the body. About midway between the coxofemoral and tibiofemoral joints there appears a line extending from the medial surface to a point about half way to the lateral edge of the femur (as if the femur were fractured), which becomes apparent

when a shadow is cast upon the femur (Plate 7,L). Sometimes this line appears closer to the trochanter. Other legs may also exhibit this fracture line (Sokoloff, 1964h).

(5) Reduced tibia (*rt*, Sokoloff and Hofer, 1964). Found in the same population as "akimbo," this mutation has proved to be allelic with *dfl* (linkage group IV). It has stronger expression: the tibia may be completely missing with the tarsi originating from the femur, or it may be present in reduced form: sometimes it is a short, curved tibia, in other cases it is hemispherical in shape; in still others the tibia is drawn into a very slender, almost filamentous shape connecting the tarsus and femur (Plate 4,I). If so, the legs may break off shortly after eclosion. The tarsus is generally complete, but sometimes the first segment is firmly fused with the tibia. Sometimes the lateral surface of the tibia is extended posteriorly into a hooklike structure more or less paralleling the femur, and in some individuals the tibial spurs develop more proximally on the tibia than normal (Plate 4, H-2, right). Because of the allelism test results, this mutant is renamed *dfl-1* (Sokoloff, 1965c).

(6) Deformed femur (*dff*; Hoy and Sokoloff, 1964). Possibly a phenodeviant found in the descendants (more than three generations) of an irradiated (12 kr) male × unirradiated female. The defect can be identified more often in the first pair of legs: the femur is bent toward the head at or near the femorotrochanteral joint, and somewhat reduced in size. The femora of the middle pair of legs may also be considerably reduced in size, and in some cases the femur is bent cephalad (Sokoloff, 1965c).

(7) Deformed tibia (*dft*, Hoy and Sokoloff, 1964). Spontaneous. Found in descendants of a strain reconstituted from four highly inbred strains marked with "sooty" (over forty generations of brother-sister mating) which did not exhibit this character. Resembles *dfl*, but tests of allelism have not been performed (Sokoloff, 1965c).

k. GENES AFFECTING THE UROGOMPHI IN THE LARVA OR THE PUPA

(1) Extra urogomphi (*eu*, Lasley and Sokoloff, 1960). Spontaneous. The effect of the gene is apparently restricted to the preimaginal stages. The larvae and pupae have one or two extra urogomphi (Plate 8,G), the supernumerary urogomphi being located between the segments bearing the genital lobes and the normal urogomphi. It is not known what morphological changes occur in the adult. Crosses of *eu* × *eu* gave 39 normal and 23 abnormal males, and 16 normal and 28 abnormal females, which suggests that the mode of inheritance is either as a recessive with

incomplete penetrance or as a dominant lethal in homozygous and semi-lethal in the heterozygous condition (Lasley and Sokoloff, 1961).

(2) Urogomphiless (*u*, Roark, 1964). Spontaneous autosomal recessive in a strain inbred for seven generations. The pupae may have one or both urogomphi missing and the genital lobes may also be deleted. Variable expression, poor viability, and penetrance reduced 65%, probably due to modifiers (report by Bell, 1964).

l. Genes Affecting the Genitalia

(1) Emasculated (*em*, Hoy and Sokoloff, 1964). A male was found in the *Fta c*/+*c* stock and another male in some crosses between *ble*+/+*sa* (in attempts to isolate a *ble sa* stock) which, upon squeezing, had no apparent aedeagus. After dissection these beetles proved to have normal testes and accessory glands, but the aedeagus had failed to evert, forming a ball-like sclerotized structure (Plate 7,P). So far no attempts have been made to determine the mode of inheritance of *em*, but by analogy with a similar mutation found in *T. confusum* (see p. 103), it probably is due to a sex-limited autosomal recessive gene (Sokoloff, 1965c).

m. Genes Affecting the Biochemistry of the Fluid in the Stink Glands

Melanotic stink glands (*msg*, Sokoloff and Hoy, 1964). In normal *Tribolium* there are two pairs of odoriferous or stink glands, one pair in the prothorax and the other in the abdomen. These glands are connected with reservoirs which are filled with a volatile liquid which is yellowish in young beetles but dark reddish-brown in old beetles. The fluid in *T. castaneum* has been found to consist of 2-ethyl-1,4-benzoquinone, 2-methyl-1,4-benzoquinone (toluquinone), and 2-methoxy-1,4-benzoquinone (Alexander and Barton, 1943; Loconti and Roth, 1953), and an oily secretion of molecular weight 179 (Roth and Stay, 1958). In *msg* the reservoirs in beetles 1 or 2 weeks old become visible through the exoskeleton owing to a change in the composition of the substances contained in the reservoirs. These substances form a solid black mass which by analogy with a phenotypically identical mutation in *T. confusum* (see below) must consist of a high molecular weight polymeric material. The black mass may assume the shape of the reservoir or may break up into several components, some appearing dotlike (Plate 12,F). As in *T. confusum* the contents in the prothoracic reservoirs undergo the chemical changes earlier than the abdominal reservoirs. Hence, the phenotype of the prothorax is more reliable for identification of *msg* than the pheno-type of the abdomen (Sokoloff, 1965c).

n. GENES HAVING PLEIOTROPIC EFFECTS

It is not unusual in *Tribolium* to find genes which have pleiotropic effects. Among the genes already identified with specific linkage groups it has been pointed out that the *pd* and *ser* sex-linked genes affect the antennae and the tarsi; *ma*, also sex-linked, affects the whole body; and the sex-linked *py* markedly reduces the body size of the beetle. In linkage group IV *Be* and *Df* affect the eye and the antenna, respectively, and both mutants exhibit recessive lethal effects in the dominant homozygote. *Mo* in group VI behaves as does *Be* in affecting the eye and in its recessive lethal effect. In linkage group VII *Fta* visibly affects the antennae, tarsi, and elytra; *Sa* and its numerous dominant alleles affect the antennae and the tibiae, while the three *sa* semidominant or incompletely recessive alleles affect the antennae and primarily the femora. *Fta* and *Sa* also have recessive lethal effects. In linkage group VIII *ap* affects the antennae, the metathorax, and the distal tarsal segments. In linkage group IX *ptl* affects the prothorax and/or the first pair of legs.

In this section are included a few more genes having a major effect on more than one body structure.

(1) Spikes on trochanters and antennae (*sta*, Sokoloff, 1963). This homeotic mutant has been found three times in this laboratory. Two of these occurrences were found about 2 years apart in crosses involving squint (Sokoloff, 1964d,h), and it is possible that the genes, owing to their incomplete penetrance, may have been overlooked. But the third occurred in descendants of an irradiated normal male mated with an unirradiated female, both kindly provided by Dr. Howard E. Erdman of the Biological Laboratory, Hanford Laboratories, General Electric Company, Richland, Washington, from a Brazil strain which was not available at this laboratory (and which also yielded the *rt*, *akb*, and *jac* mutants described above). The following description is from the well-established stock which was founded from beetles found in the course of determining the frequency of crossing-over between *ap* and *sq*.

Several males and females were found among the progeny of a single pair mating which attracted the writer's attention because the dead or dying imagoes had failed to shed the pupal skin. When the pupal skin was removed, the trochanters of all the legs and the second antennal segment exhibited very large spikelike growths (Plate 10,A). The parents of these beetles were transferred repeatedly to fresh food and they produced a number of additional *sta* progeny with less marked effects, sometimes the spikes being confined to the antennae (but not necessarily arising from the second antennamere) or to the trochanters (Sokoloff, 1964d). Mated *inter se* for several generations *sta* beetles produced but a few individuals resembling their parents. A chance examination of the larvae in the stock jar by the author's assistant M. Hoy revealed that *sta* has a distinctly recognizable effect on the larvae: the legs may exhibit fleshy growths, apparently originating from the trochanters, which may be almost as large as the whole leg (Plate 10,B–D), they may be very small or they may even be absent. The antennae may be modified in various ways, sometimes resembling the antennae of prothetelous individuals, and occasionally being completed duplicated. While the incidence of abnormal larvae now approaches about 80% in the stock, the incidence of imagoes drops to a very low level. The leg growths apparently are resorbed at metamorphosis, and imagoes emerging from larvae known to have leg or antennal abnormalities may appear normal. The range of expression of the antennae in affected *sta* adults has become quite variable: spikes may originate from different antennameres even in one individual. The spikes may be small and pointed, or very long, and filamentous. Some spikes may exhibit various degrees of segmentation. In many individuals instead of spikes there may be segmented branches arising at any point of the antenna, or the antenna may be completely duplicated. Examples of these various phenotypes are shown in Plate 10, E–O. Hoy has also noted a great increase in prothetelous individuals in the *sta* stock. At the present time it is not possible to state whether this is a definite characteristic of the *sta* strain, or whether these were induced by environmental conditions. Flour beetles at critical stages of development, exposed to quinones released by imagoes or to other gases (Roth and Howland, 1941), have been seen to undergo premature metamorphosis of antennae, legs, and both pair of wings. These larvae-pupae die before becoming adults. Further observations on this phenomenon are in progress.

It is clear that some of the *sta* beetles, judging by the appearance of the antennae, resemble the *bra* phenodeviant described by Dawson (compare with Plate 10,P). Unfortunately *bra* was no longer available, having been lost owing to infertility following intensive inbreeding and selection. Hence, it was not possible to compare *bra* with *sta* larvae, or to perform tests of allelism. A stock was available of the earlier *sta*-like mutation found in this laboratory in other crosses involving *sq*. When first found, this mutant had a small process arising from the second

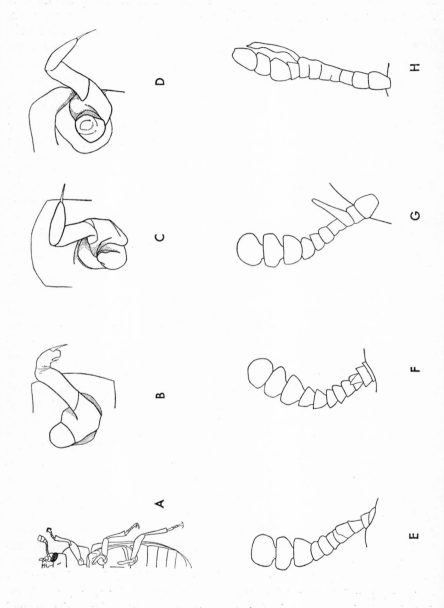

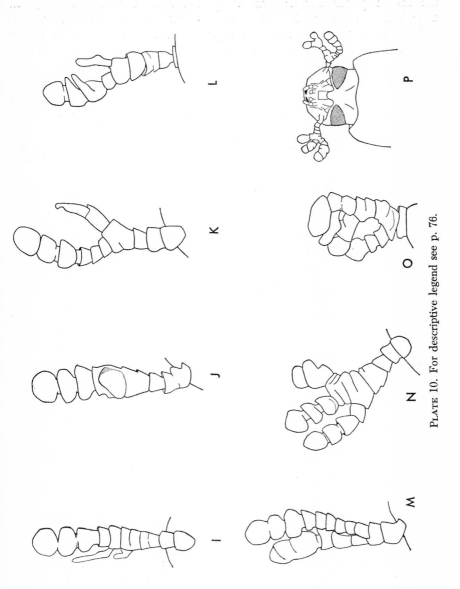

PLATE 10. For descriptive legend see p. 76.

antennal segment and it was designated horned antennae (*ha*). In crosses carried out by R. Paige, one F_3 was obtained with fused tarsi and the tibia of the left middle leg swollen and segments 3–5 of the right antennae fused. In another creamer a male was found with segments 2–3 of both antennae fused; the tibia of the left middle leg had a spikelike branch and the left hind leg was badly deformed. In other F_3 creamers were found rare individuals with branched antennae. Tests of allelism between *ha* and *sta* revealed that they are the same gene, and thus *ha* is renamed *sta-1*.

In *sta-1* some individuals possess supernumerary growths in the tibiae, while in *sta* these growths arise from the trochanter (Sokoloff, 1965c).

In Balazuc's (1948) monograph on the teratology of Coleoptera the following species of beetles are cited and illustrated whose antennae are two-branched or three-branched: *Ibidion* sp., *Dihammus musivus, Rhamnusium bicolor, Magdalis barbicornis, Collops tricolor, Toxospathius inconstans, Zabrus ovalis, Arthrodactylus elongatus, Lucanus cervus, Julodis Clouei, Cantharis pellucida, Carabus obsoletus* var. *euchromus, Macrotoma* sp. *prope senegalensis, Neodocardion egregium,* and *Odontolabis Stevensi.* Jayne (1880) reported a *Prionus californicus* with all three pair of legs duplicated entirely beyond the femur (cited and illustrated in Balazuc, 1948, p. 170). Balazuc (1948, p. 172) himself described and figured a *Lagria hirta* which in addition to fusions of segments 3–4 and 5–6 of the right antenna, had spikelike processes arising from the femur of all three legs, and from segment 8 of the left and from segment 10 of the right antenna. I believe that the *sta* mutation provides a genetic explanation for such monsters as have been cited in the teratological literature.

(2) Troll (*tro,* Hoy and Sokoloff, 1964). Eight pupae of both sexes of *tro* have been isolated, suggesting an autosomal mode of inheritance. The adults emerging from the pupae resemble the sex-linked *ma* (see Plate 2,B) in every respect except that the tarsi of all the legs are reduced in number by one or two tarsomeres (Plate 9,D,E). Viability is

---

PLATE 10. "Spikes on trochanters and antennae" (*sta*) and "branched" (*bra*) in *Tribolium castaneum.* (A) Adult beetle showing spikes arising from trochanters in all three pairs of legs. (B–D) Left legs of an *sta* larva showing processes medial to the legs. These processes are usually resorbed before the adult forms, but may show up in the adult as in A above. (E–O) The expression of *sta* in various adults. Note that one or more processes may arise in any segment (F–L). Also, the *sta* gene may cause two or more segmented branches to form (M–O). (P) The appearance of the antennae in the phenodeviant *bra.*

greatly reduced. It is not possible to state at present whether the gene is lethal or semilethal (Sokoloff, 1965c).

(3) Antennae and tarsi fused (*atf*, Sokoloff, 1961). Spontaneous recessive of variable expression found in the course of determining linkage relationships between *aa* and c^s. The *atf* gene causes a fusion of variable segments of the tarsi and sometimes the effect extends to the funicular segments of the antenna, but these appendages may be free of fusions while the tarsal segments are nearly always fused. The data given in Table 13 are confined to the distal segments of the tarsus. After the legs had been mounted it became evident that when *atf* is strongly expressed there is also a partial fusion of the first tarsomeres (Plate 1,F 1–3) which is not very obvious when the tarsi are examined *in situ* (Sokoloff, 1965c).

(4) Sleek (*slk*, Sokoloff and Hofer, 1964). This mutant appeared

TABLE 13

Tarsal fusions in *atf* in *Tribolium castaneum*

	Males						Females					
	Leg 1		Leg 2		Leg 3		Leg 1		Leg 2		Leg 3	
Number	R	L	R	L	R	L	R	L	R	L	R	L
1	4–5	4–5	4–5	4–5	3–4	3–4	4–5	4–5	4–5	4–5	0	0
2	4–5	4–5	4–5	4–5	2–4	2–4	4–5	0	0	4–5	3–4	3–4
3	4–5	4–5	4–5	4–5	3–4	3–4	4–5	4–5	3–5	3–5	1–3	1–3
4	4–5	4–5	4–5	4–5	3–4	3–4	4–5	4–5	0	4–5	2–4	2–4
5	4–5	4–5	4–5	4–5	3–4	3–4	4–5	4–5	3–5	3–5	3–4	3–4
6	4–5	4–5	4–5	4–5	2–4	3–4	4–5	4–5	0	4–5	3–4	3–4
7	4–5	4–5	4–5	4–5	2–3	2–3	4–5	4–5	4–5	4–5	3–4	3–4
8	4–5	4–5	4–5	4–5	2–4	3–4	4–5	4–5	3–5	3–5	2–4	2–4
9	4–5	4–5	4–5	0	1–4	2–4	4–5	4–5	3–5	3–5	1–3	1–4
10	4–5	4–5	4–5	4–5	2–4	2–4	4–5	4–5	4–5	4–5	3–4	1–3
11	3–5	4–5	—	3–5	1–4	1–4	3–5	3–5	3–5	3–5	1–4	1–3
12	4–5	4–5	4–5	4–5	3–4	3–4	3–5	3–5	3–5	3–5	2–4	2–4
13	4–5	4–5	3–5	3–5	1–4	1–4	4–5	4–5	3–5	3–5	2–4	2–4
14	0	4–5	4–5	0	0	3–4	4–5	4–5	4–5	4–5	2–4	2–4
15	4–5	4–5	4–5	4–5	2–4	3–4	4–5	4–5	4–5	4–5	2–4	2–4
16	4–5	4–5	4–5	4–5	2–4	3–4	4–5	4–5	4–5	4–5	2–4	3–4
17	3–5	4–5	3–5	3–5	1–4	1–4	4–5	4–5	2–5	2–5	2–4	2–4
18	3–5	3–5	3–5	3–5	2–4	2–4	4–5	4–5	3–5	3–5	2–4	2–4
19	3–5	3–5	3–5	3–5	2–4	2–4	4–5	4–5	4–5	4–5	3–4	3–4
20	4–5	4–5	3–5	3–5	2–4	2–4	4–5	4–5	0	0	2–4	3–4

spontaneously in crosses attempting to establish linkage between *bal,* *lod,* and *p.* Two female imagoes were produced from a single-pair mating and they died without leaving progeny. Hence, the mode of inheritance has not been established. However, the two females had identical appearance (Plate 9,F) so that there is no doubt that the following abnormalities, involving some major taxonomic characters, had genetic basis:

(*a*) Antennae: first and second basal segments not fused but somewhat reduced in size; segments 3–8 of the funicle and 9–11 of the club fused into a continuous paddlelike structure exhibiting no segmentation.

(*b*) Head rounded and somewhat reduced in size.

(*c*) Prothorax also rounded, with the anterolateral extensions missing, and more convex dorsally. Ventrally the epimeron (E), trochantin (T), and the medial extension of the episternum between E and T were missing so that the coxae of the first pair of legs were exposed, i.e., they were "open behind" in taxonomic terms. The sternellum was reduced in size.

(*d*) Mesosternum reduced in size and the sternellum short and pointed, failing to separate the middle coxae.

(*e*) Metasternum without median anterior projections and with the median groove missing. The hind coxae incomplete, with a circular opening showing the proximal end of the femur lying within.

(*f*) Elytra and membranous wings abbreviated, reaching only the posterior edge of the first apparent abdominal segment.

(*g*) Medial anterior projection of the first apparent abdominal segment missing or vestigial. The remaining abdominal segments were separated by a wide membranous (unsclerotized) area.

(*h*) The last apparent abdominal segment was considerably reduced in size and less rounded.

(*i*) Proximal podomeres normal, but the tarsus conforms to the formula 4–4–3, with partial or complete segmentation visible between all adjacent remaining tarsomeres (Sokoloff, 1965c).

4. Linkage Maps in *Tribolium castaneum*

The linkage relationships of various genes has been determined. They are summarized in Fig. 1. Gene symbols given in parenthesis have been identified with that particular linkage group, but their map position is yet to be ascertained from three-point crosses. The maps are based on the data given below. In certain cases the data have not yet been published. In other cases, indicated in brackets, the experiments need to be or are being, repeated, particularly for certain genes which have incom-

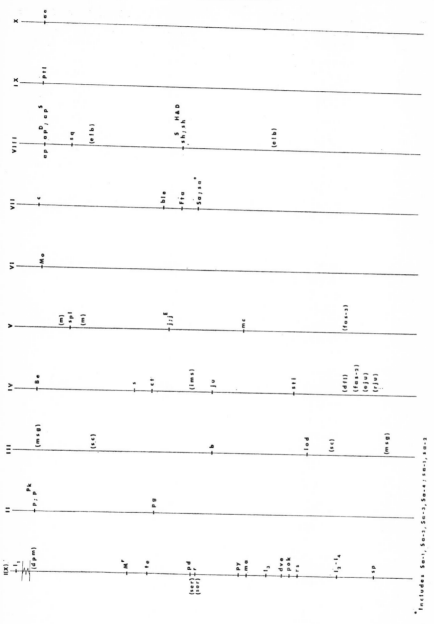

Fig. 1. Linkage maps in *Tribolium castaneum*.

*Includes Sa-1, Sa-2, Sa-3, Sa-4; sa-1, sa-2

plete penetrance or marked reduction in viability, and for which the
data appear inconsistent. More complete data can be found in the series
of papers entitled "Linkage studies in *T. castaneum*" by the writer and
associates (see bibliography) and in Dawson (1964c).

Frequency of Crossing-Over (in per cent) between Genes in Various Linkage Groups in
Tribolium castaneum

I	IV	VII
$pd-l_1 = 61$	$Be-s = 25$	$ble-c = 42$–44
$pd-dpm = 40$	$s-ims = 11$–23	$ble-Fta = 3$–6
$r-M^r = 16$	$Be-ims = 43$	$ble-Sa = 4$
$r-te = 11$	$Be-ju = 44$	$Fta-sa = 18$
$pd-r = 1$	$Be-ct = 29$	$Fta-c = 34$–45
$pd-ser = 0$	$ju-ct = 7$	$sa-c = 38$
$ser-r = \, < 1$	$s-sti = 41$	$Sa-1-c = 40$
$pd-ma = 13$	$sti-ims = 35$	$Sa-sa = 0$
$ma-py = 2$	$Be-dfl = 27$	VIII
$py-l_3 = 5$	$s-dfl = 24$	$elb-sh^S = 23$
$py-dve = 11$	$[dfl-ims = 50]$	$sh^S-sq = 28$
$py-rs = 15^*$	$[Be-fas-2 = 27]$	$ap^D-sq = 7$
$py-l_8 = 23$	$[fas-2-s = 25]$	$ap-ap^D = 0$
$py-l_4 = 25$	$[rju-s = 13]$	$sh^S-sh^{H \, and \, D} = 0$
$pd-sp = 46$	$[rju-fas-2 = 38]$	
$pok-dve = 2$	$[eju-fas-2 = 38]$	
II	**V**	**IX**
$p-pg = 30$	$j-spl = 29$	ptl
$p-p^{Pk} = 0$	$j-mc = 25$	
III	$spl-mc = 41$	
	$m-j^E = 22$†	**X**
$b-lod = 24$	$m-mc = 40$†	aa
$b-msg = 44$	$j^E-mc = 18$†	
$b-sc = 28$–33		
	VI	
	Mo	

* Data of Reynolds (1964).
† Data of Eddleman (1964).

B. *Tribolium confusum*

1. Sex-Linked Genes

a. VISIBLES

(1) Striped (*St*, McDonald, 1959), the first sex-linked mutation found
in *T. confusum*, is a dominant with recessive lethal effects in the female

and effects lethal or nearly so in the male. The few sterile males which survive to the imago have whitish elytra and die at an early age. $St/+$ young females have a broad, pigmentless stripe in the middle of each elytron, (Plate 11,A) but on aging this becomes pigmented and generally indistinguishable from wild type. The heterozygous females have good viability (McDonald, 1959 a,b).

(2) Eyespot (*es*, McDonald and Peer, 1961), is a sex-linked recessive mutation affecting eye color. The mutant phenotype is readily recognized in the larva by the absence of the black ocelli, and in the pupa by the development of reddish ommatidia. It is recognized with more difficulty in the adult: the eye requires strong illumination and careful scanning, which is facilitated by intercepting the light with some object and casting a shadow over the eye. Then the reddish phenotype becomes visible in a small area as indicated in Plate 12,A. Otherwise these beetles would be misclassified as wild type. McDonald and Peer (1961b) have determined that *es* is about 38 units away from *St*.

A lighter, recessive allele of *es* (es^{lt}, Sokoloff, 1964) has been found. This allele is readily identifiable in any stage including the adult. Excepting for the marginal ommatidia, which remain black because of the underlying black-pigmented ocular diaphragm, the central ommatidia appear reddish to pink throughout the eye (Sokoloff, 1964h).

(3) Labiopedia (*lp*, Sokoloff and St. Hilaire, 1962), is a spontaneous sex-linked recessive mutation of somewhat variable expression but complete penetrance found as a single male pupa (Sokoloff, 1963b,e). It is noteworthy because it is the second homeotic mutation found in *Tribolium* (see antennapedia in *T. castaneum*) but the first sex-linked mutation of this type among the Insecta. Furthermore, in modifying the labial palps to walking legs it has been possible to compare its effect with that of proboscipedia in *Drosophila* (Bridges and Dobzhansky, 1933), which also affects the mouthparts. The anatomical details in the normal and labiopedia larva, pupa, and adult, have been described elsewhere (Daly and Sokoloff, 1965). For the present review, it may be sufficient to point out that, as in other mutations affecting the appendages in *Tribolium*, it happens that the effect of the mutant gene becomes noticeable as soon as the larva hatches from the egg. As can be seen in Plate 11, in the *lp* larva the labial palps are replaced by complete larval legs (Plate 11,H); the *lp* labial legs are visible in the pupa (Plate 11,I) and the adult (Plate 11,E) emerges with a pair of labial legs attached to an oversized labium. The labial legs usually consist of the claw-bearing tarsus, tibia, femur, trochanter, and part of the coxa. They possess certain muscles, but they do not appear to move *in situ*, although legs mounted in saline

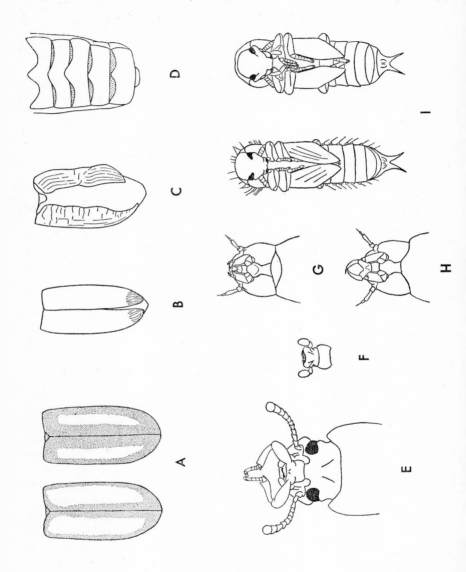

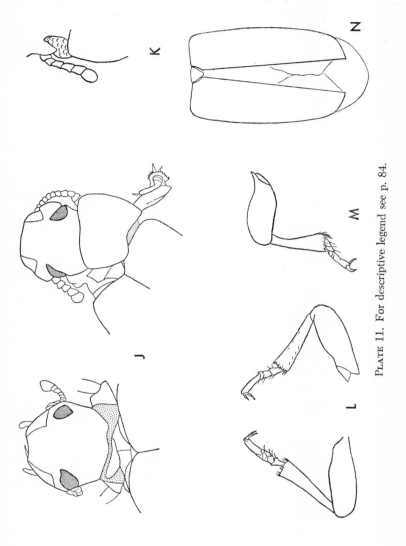

PLATE 11. For descriptive legend see p. 84.

are seen to twitch. These labial legs, which impede the feeding activities of the beetle, are often broken off by the adults themselves (by autotomy), with the help of the forelegs. But even if both legs are broken off at the base, the mutant is recognizable by the much larger labium (which is about twice as wide and long as the normal). In intact beetles there is some variability in expression, but in general, a recognizable leg is formed in symmetrical fashion. Occasionally one can observe beetles with extremely small heads covered with several layers of larval skin, and pupal skin. These beetles die at an early age without reproducing, but when the preimaginal skins are dissected out they have been found to be *lp*. Other remarks will be postponed until later, in the section on homeotic mutants.

Mutant *lp* is located about 47 units from *es* away from *St* (Sokoloff, 1964e). Although it behaves as semilethal to lethal on out-crossing it is (so far) one of the most useful markers for the X-chromosome of *T. confusum*.

(4) Pointed abdominal segments (*pas*, Sokoloff and Ho, 1963). This homeotic mutation, which exhibits serial homology of the abdominal segments, was found as a single male in experiments designed to determine the genetic load in *T. confusum*. The male died without leaving progeny, but two of eleven female sibs proved to be heterozygous for the gene. The *pas* gene is a recessive with fairly good viability, but males, although they may live for as long as 2 months, are sterile. The sterility is probably a mechanical one, since upon dissection normal internal and external reproductive organs, including a normal aedeagus, are found. As can be seen in Plate 11,D, the apparent abdominal segments 2–5 (which are similar to each other but distinct from the apparent first abdominal segment in normal beetles, Plate 12,C,E), are modified in phenotype and closely resemble the apparent first abdominal segment, acquiring

---

PLATE 11. Sex-linked mutants in *Tribolium confusum*. (A) Two examples of Striped (*St*). (B) Thickened elytral tips (*tet*). (C) Crumpled (*cru*). (D) Pointed abdominal sternites (*pas*). Note that all segments appear like the first apparent abdominal segment. (E) Adult labiopedia (*lp*). (F) The labium of a normal adult with the three-segmented labial palps, enlarged about two times the enlargement in E. (G) Ventral view of the normal larval head. (H) Ventral view of an *lp* larva. (I) Normal pupa and an *lp* pupa. Note the extra legs emerging from under the head in *lp*. (J) Two examples of prothoraxless-like (*ptll*). (K) The antenna in "antennae and elytra reduced" (*aer*). (L,M). The first pair of legs and one of the hind legs to show the effect on the tarsi in *aer*. (N) The appearance of the elytra in *aer*.

a medial anterior projection on each sternite. With this change in form the sternites no longer overlap effectively, the spaces between them, lateral to the medial projections, becoming unsclerotized. Also, the posterior end of the apparent fifth sternite is not rounded as in the normal beetles, but part of this area appears to be missing, the abdomen acquiring a more angular form.

The *pas* gene is located between *es* and *lp* about 3 units from *lp* (Sokoloff, 1965b).

(5) Miniature appendaged (*ma*, Sokoloff and Bywaters, 1961). If proper allowance is made for the difference between the normal phenotype of *Tribolium castaneum* (*CS*) and *T. confusum* (*CF*), this mutation looks identical in phenotype with *ma* in CS, (Plate 2,B,D), the pleiotropic effect of *ma* in *CF* extending throughout the body, but being identifiable more readily by examining the legs (which become short and thick) and the elytra (which are reduced to about two-thirds of the normal length). While there is no question that *ma* was sex-linked, its position in the X relative to other genes is dependent upon its rediscovery since through a technical error the mutant was lost shortly after it was found (Sokoloff, 1961a).

(6) Antennae and elytra reduced (*aer*, Sokoloff, 1962). Found among the progeny of a single female derived from an Oakland, California, flour mill. The elytra and membranous wings in the pupa are shorter by about one abdominal segment exposing the tarsi of the hind legs. In the adult the elytra are short and split, the antennae are short, consisting of an average of seven segments. The tarsal segments are variously fused (Plate 11,K–N). This is a recessive lethal since the few surviving male imagoes die shortly after eclosion and before they are able to mate (Sokoloff, 1962d). Preliminary crosses suggest *aer* is located about 24 units from *lp*, away from *es*, (Sokoloff, 1965b). These linkage studies showed the following peculiarities: (1) from the mating *es lp* male with females derived from the stock where *aer* was being maintained, female virgins were chosen from those matings showing *aer* among the progeny. These females, therefore, were either ++*aer*/*lp* *es* + or ++/*lp* *es*. Out of sixteen females selected only one was of the latter genotype, although eight were expected to be so. (2) the females were mated to *es* males. Their progeny showed a significant shortage of *es*/*es* females as contrasted with wild-type females. Since an analogous excess of lethal-carriers was observed while mapping the position of lethal-3 (*l₃*) in *T. castaneum* (see above), these phenomena need to be investigated further.

(7) Red (*r*, Sokoloff, 1962). A recessive found in a "natural" population collected in an Oakland, California, flour mill. Its phenotype varies as a function of age: it is readily identified in the pupa and young imago by the reddish ommatidia, but it darkens in older beetles. Tests of allelism between *r* and *es* indicate they are not allelic. The *r* gene has been located at 48 units from *lp*, but at this writing it is not known whether it is the vicinity of *es* or in the opposite direction (Sokoloff, 1964e).

(8) Thickened elytral tips (*tet*, Sokoloff, 1963). A recessive found in crosses attempting to establish the location of *St, es,* and *lp*. The elytral tips appear somewhat rounded and raised. In its extreme expression a tiny blister develops at the tips of the elytra (Plate 11,B). Penetrance is probably complete, but the mutant is difficult to identify since in mild expression it resembles the normal phenotype. Preliminary crosses locate this gene 53 units from *es*, but three-point crosses have not been performed for a more precise location (Sokoloff, 1964h).

(9) Crumpled (*cru*, Hoy and Sokoloff, 1964). A recessive, characterized by the possession of split, drooping elytra, somewhat reduced in size in the adult so that the posterior tips cover only part of the posterior abdominal segment (Plate 11,C). The dorsal surface of one or both elytra may present a roughened, sometimes wavy appearance. In addition many of the beetles exhibit a huge blister in the proximal portions of the elytra. Detectable in the pupa either by the fact that the elytra do not reach the tips of the hind tarsi, or by the presence of a blister. The latter pupae resemble "akimbo" in *T. castaneum* or *thu,* and *thu*[s] in *T. confusum*. Since the mutant was just discovered, information on penetrance or viability is not available (Sokoloff, 1965c).

(10) Prothoraxless-like (*ptll*, Sokoloff and Daly, 1964). Twenty *ptll* male specimens, progeny of a single female, were recently found. Their phenotype is similar to *ptl* in *T. castaneum* in the following ways:

In the most weakly expressed beetles the prothorax is symmetrical and reduced by about one-fifth (along the longitudinal axis of the body), the reduction appearing to take place in the posterior part of the prothorax. In more strongly affected beetles the pronotum from the left or right side of the body is missing, the remaining portion acquiring a triangular shape. In an attempt to cover both halves of the prothorax, this triangular piece may be slightly modified in direction (Plate 11,J). On the ventral portion of the prothorax the pro- and basisternum may be considerably reduced in size and deformed, but the posterior portion of the basisternum and the sternellum generally are unaffected. The legs may be deformed, the deformity resembling that in the *T. castaneum ptl* heterozygotes, but so far a reduction of the forelegs to the extent that they become vestigial has not been observed, perhaps because *ptll/ptll* beetles are not yet available.

The *ptll* mutants differs from *ptl* in that the labium in the latter is not affected, but in *ptll* the labium appears considerably reduced in size, and oriented ventrally. The labial palps are three-segmented but likewise reduced in size. In one specimen the gular region was found fused with the prothorax. In mildly deformed beetles the head retains the normal orientation of the head characteristics of prognathous beetles, although the head may appear broader than normal, and the head may be twisted at an angle. In strongly expressed *ptll*, where the prothorax is badly deformed, the prothorax is oriented dorsally at an angle of about 45° from the body and the head appears at a higher level than, but parallel to, the longitudinal axis of the body.

The mutant is readily recognizable in the larva or the pupa by deformities in the prothorax. Imagoes have no trouble eclosing from the pupa, but the majority survive only briefly after becoming imagoes, probably because they are unable to feed (Sokoloff, 1965c).

b. LETHAL

Lethal-1 (*l₁*, Sokoloff, 1962). Spontaneous recessive found in crosses involving *rus* (Sokoloff, 1962d). Located 40 units from *es*, but no other information available in regard to its map position since at the time it was tested other sex-linked genes, surviving as males, were not available (Sokoloff, 1964e).

2. Autosomal Genes

a. GENES AFFECTING EYE PIGMENTATION

Aside from the sex-linked *es*, *es^lt^*, and *r* genes, the following eye color mutants are known:

(1) Pearl (*p*) (Graham, 1957, *p^H^*, and *p^H^-1* reported from the Wantage and Malta strains, respectively, by R. W. Howe, Pest Infestation Laboratory, Slough, England, 1962, and *p^s^*, Sokoloff, 1963). This recessive gene blocks the formation of pigment in the larval ocelli and in the ommatidia, but as in other tenebrionids the eye appears spectacled or bicolored (Plate 12,B) because under the marginal ommatidia lies the ocular diaphragm which is pigmented black and which is readily seen in all eye color mutants producing light red or crystalline omma-

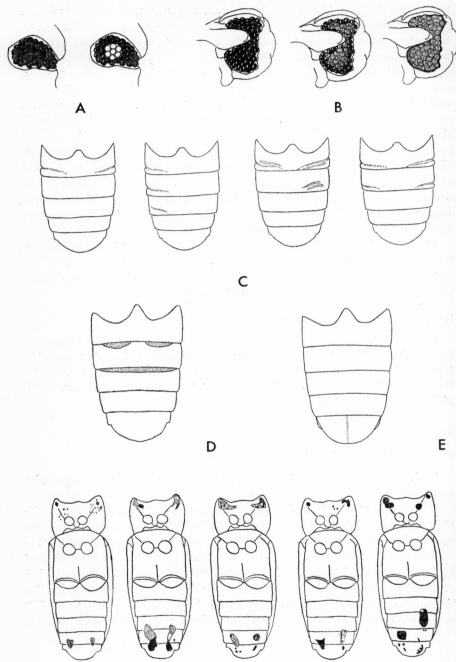

tidia. The *p* gene prevents the formation of pigment in the Malpighian tubules (Shaw, 1965).

Because *p* in *T. confusum* is epistatic to other eye color genes; because it appears to be unstable judging by the fact that at least four independent occurrences of the mutant have been observed, and because beetles heterozygous for pearl undergo somatic mutations resulting in individuals with one eye black, the other pearl, it has been suggested that *p* in this species must be homologous with *p* in *T. castaneum*. For these various reasons *p* has been suggested as a useful marker for identifying the second linkage group for all tenebrionids (Sokoloff, 1962b).

(2) Dirty pearl eye (*dpe*, Sokoloff, 1963). This recessive mutation produces a reddish eye resembling *es* and *rus* in the imago. In the pupa and teneral adult the eyes in *dpe* appear pearl-like except that the central ommatidia appear "dirty." On aging, the eyes of the imago become dark red, perhaps darker than *rus*, resembling the chestnut mutation in *T. castaneum*. Mutant *dpe* is located about 5 units from *p*, and about 30 units from *cas*, the order being *p–dpe–cas* (Sokoloff, 1964f).

(3) Ruby (*rby*, Sokoloff, 1964) is a recessive, dark red-eyed mutant perhaps darker than *rus*, approaching a Burgundy red in aged beetles. The phenotype of the latter is difficult to identify except by intense illumination and shading of the eye (Sokoloff, 1964f).

(4) Ruby spot (*rus*, Sokoloff, 1961) is a recessive, resembling *es* in phenotype and appearing as in Plate 12,A, the light circled area representing the amount of dark red color which can be observed as the ventral aspect of the eye is examined. This gene is linked with, and about 42 units from, *b* (black) (Sokoloff, 1964f).

(5) Frosted (*fro*, Hoy and Sokoloff, 1964), is a recessive mutation not allelic to pearl, but which nevertheless resembles the phenotype of this mutation (Sokoloff, 1965c).

PLATE 12. Various autosomal mutants in *Tribolium confusum*. (A) Ventral view of the normal eye, and approximate size of pigmented area in such mutants as eyespot (*es*) and ruby spot (*rus*). (B) Lateral view of the normal, pearl (*p*), and pearl with light ocular diaphragm (*p, lod*) eye. Note that in *p, lod* the ommatidia are uniform in color, in *p* the ocular diaphragm is visible as a dark ring under the ommatidia. In the normal fully aged beetle the ommatidia are uniformly black and the ocular diaphragm is not distinguishable. (C) Four examples of "creased abdominal sternites" (*cas*). (D) The sternites incomplete (*sti*) phenotype. (E) Median abdominal groove (*mag*). (F) In melanotic stink glands (*msg*) the gland reservoirs become visible through the exoskeleton owing to the formation of a black polymeric material of high molecular weight. (All drawings except D and E by permission of the Editor, *Canadian Journal of Genetics and Cytology*.)

(6) Light ocular diaphragm (*lod,* Sokoloff, 1962). Although strictly speaking this is not an eye color mutation, this recessive is included here because it can be detected only when the normally black pigment of the ommatidia fails to form as a result of an eye color gene. The *lod* gene prevents the formation of melanin from the ocular diaphragm so that eye color mutants no longer appear "spectacled," but the ommatidia become uniform in expression, the color depending on the eye color gene involved (Plate 12,B). In contrast to the *lod* gene in *T. castenum, lod* in *T. confusum* segregates independently from *b.* Conceivably *lod* may still be in the linkage group marked with *b,* but far away from it. If so, this can be determined only with the aid of gene(s) intermediate in position between *b* and *lod* (Sokoloff, 1964).

b. Genes Affecting the Morphology of the Eye

In *T. castaneum* five mutants (*Be, gl, mc, Mo,* and *sq*) are known to affect the eye in various ways. *Mo* and *mc,* in addition, reduce the size of the head behind the genae. So far no genes have been discovered in *T. confusum* producing comparable abnormalities. Occasionally, in matings involving *lp,* imagoes have been recovered whose heads are covered with larval and pupal skins. When these are removed, a beetle with a tiny head is seen. The antennae, the mouthparts, and the eyes are proportionately reduced. These beetles have failed to produce progeny, but it is believed that this condition is the result of the labiopedia gene. In some stocks not involving *lp* some larvae have been found whose heads are only about half as large as those of their normal sibs. When these are allowed to develop to imagoes and bred, their progeny appear to have smaller heads, but there is much overlap and critical measurements are necessary before the mode of inheritance of this condition can be established.

c. Genes Affecting Body Size

Tiny (*ty,* Sokoloff, 1961). Small sized beetles have often appeared in small numbers in various crosses in *T. confusum,* some approaching the size of pygmy (*py*) in *T. castaneum.* Because such a mutation would be very useful in mapping sex-linked genes, considerable time and effort have been spent in determining whether these small beetles result from the action of a single sex-linked recessive or semidominant gene. The efforts have proved fruitless, and the reduction in size has been attributed to the action of a large number of multiple factors controlling body size, grouped under the same designation as *ty* in *T. castaneum* (Sokoloff, 1961, unpublished).

d. Genes Affecting Body Color

Apart from the sex-linked *St* gene which prevents deposition of pigment in the elytra of teneral adults the following mutants have been discovered which considerably modify the red-rust or chestnut color characteristic of wild-type *T. confusum*.

(1) Black (*b*). Stanley and Slatis (1955) called this mutation McGill black and the wild type McGill Red (Stanley and Slatis, 1955). Since this practice may lead to confusion, the word "McGill" has been eliminated from the designation of "McGill black" and "McGill Red" and the latter is referred to simply as normal or wild type, for after all, these beetles are no different in body color from beetles which one can collect under "natural" conditions.

The *b* gene is a semidominant producing a body color, which more nearly matches the following color standards in Plate L of Ridgway's (1912) manual: "dull violet black" (61''''. VR-V) or "aniline black" (69''''. RVR). Hence, it resembles *b* in *T. castaneum*. The heterozygote is distinguishable from *b/b* and $+/+$, and is commonly referred to as bronze, but it appears nearest the color called "burnt umber" on Plate XXVIII (9''. OR-O.m).

This mutant has been recovered several times in unrelated stocks: Dr. D. J. McDonald, Dickinson College, Carlisle, Pennsylvania, in his population studies, has used a strain obtained from Dr. S. G. Smith at Forest Insect Laboratory, Sault Ste. Marie, Ontario, Canada (1964, personal communication). A black strain was reported by Dr. H. C. Chiang, University of Minnesota, to have occurred spontaneously in a wild strain of *T. confusum* obtained in St. Paul, Minnesota. These new recurrences of *b* might be termed *b-1* and *b-2*, respectively, since they proved to be allelic (Sokoloff, 1964h). Recently Dr. J. V. Slater, Donner Laboratory, University of California, Berkeley, has provided the writer with a black strain which appeared in a wild-type stock derived by Dr. Garth Kennington, University of Wyoming, from the Chicago strain. Superficially this new black appears identical with *b*. On the basis of tests of allelism just completed, this black is allelic to *b* and is symbolized *b-3*. It is interesting to note that *b-3* $\times+$ crosses give wild-type F_1 (the *b* $\times+$ crosses give bronze F_1), but *b-3* $\times b$ gives a black phenotype.

The *b* derived from the McGill wild-type strain for a long time after its discovery behaved differently from the other blacks in *T. castaneum* or *T. confusum:* *b* females, when mated with *b* males from the same strain produced abundant progeny. The same females mated to unrelated non-*b* males produced many eggs, but they proved to be sterile. The *b* males successfully fertilize females from any source (Stanley, 1961b). The writer was able to corroborate this observation in some early crosses involving *b, lod,* and *p.* Recently, however, neither females derived from the original *b* strain kept at McGill nor those derived from a substrain kept in this laboratory have shown any signs of sterility when they are outcrossed.

The *b* gene segregates independently from *lod.* Hence, it cannot be homologous (at least in position) with the *b* gene in *T. castaneum.* It is not allelic nor linked with *e;* it is in a different linkage group than *p.* It is linked with *rus* and *msg,* these genes being about 41 and 42 units away from *b.* Some crosses suggest *sp* may be linked with *b,* but over 43 units apart.

(2) Ebony (*e* Park, Ginsburg and Horowitz, 1945) modifies the reddish body pigment to one appearing black to the unaided eye, but under the dissecting microscope it is clearly different and more reddish than *b.* Comparisons with Ridgway's (1912) color standards, lead to the conclusion that *e* approximates "maroon" (3.0-R.m) given on Plate I or "diamine brown" (3'.0-R.m) on Plate XIII. An occasional individual may appear like "aniline black" (69''''. RV-R.m) on Plate L, but this may be a function of age.

The *e* gene is a recessive. Fertility and rate of larval and pupal development are not affected, but fecundity of *e/e* females is about 10% lower than that of normal females (Park *et al.,* 1945).

Park and associates (1945) attempted to establish the chemical nature of the difference in pigment between normal *T. castaneum* and *T. confusum* which appear the same to the eye and ebony which is clearly different. The slopes of the curves obtained from the absorption spectra of the melanin solutions prepared from normal *T. castaneum* and *T. confusum* were different, and Park *et al.* (1945) conclude that these differences may be due to "qualitatively different molecular structures or to disparities in the physical state of the melanin in these two species." On the other hand, the slopes of the curves from the melanin solutions prepared from normal *T. confusum* and ebony were identical but the mutant contained about 17.5% more melanin than the wild type.

Lerner and Ho (1960) found a spontaneous black body mutant in a strain of *T. confusum* synthesized from a number of wild-type laboratory and "naturally" occurring strains. This proved to be allelic with *e* (Sokoloff, 1962d) and it was symbolized $e^{L \text{ and } H}$.

Comparisons of *e* with the mutants sooty (*s*) and jet (*j*) in *T. castaneum* by several individuals in our laboratory led to the conclusion that *e* resembles *j* more than *s*. Since *e* is not linked with *b* (Stanley and Slatis, 1955); since *e* and e_2 (which is linked with pearl) segregate independently (Dyte and Blackman, 1962a); and since *b* and *p* are in separate linkage groups (Sokoloff, 1964c), it has been suggested (Sokoloff, 1962b) that *e* be used as a marker for linkage group V in *T. confusum*.

(3) Ebony-2 (e_2, Dyte and Blackman, 1962), is a recessive. The phenotype of this mutant is identical with the semidominant *b*, but tests of allelism performed by Stanley (Dyte and Blackman, 1962a) showed that they were not allelic. Ebony-2 was found in the second linkage group and about 2.5 units away from *p* (Dyte and Blackman, 1962a).

e. GENES AFFECTING THE PROTHORAX

Aside from the recently discovered sex-linked recessive lethal or semilethal mutation prothoraxless-like, three interesting autosomal mutants affecting the prothorax have recently been discovered:

(1) Knobby prothorax (*knp*, Sokoloff and Hoy, 1964). An autosomal recessive homeotic mutation found in a *b rus* (*sp*) selection creamer. In the pupa (Plate 13,A) the posterior lateral corners of the prothorax are extended into a growth resembling the elytral and membranous wing bud of certain vestigial mutations. As the pupa ages, this bud may become filled with fluid and in certain cases the bud may subsequently necrotize, falling off at metamorphosis. If the tissue does not necrotize, the posterior angles of the prothorax are drawn into knoblike structures, sclerotizing normally (Plate 13,B). No linkage information. Preliminary crosses suggest viability of *knp* is low (Sokoloff, 1965c).

(2) Separated epimera (*sep*, Sokoloff and Hoy, 1964). Spontaneous. Preliminary crosses suggest the gene is an autosomal semidominant with variable expression and probably incomplete penetrance. In the normal beetle the basisternum continues posteriorly between the coxae of the first pair of legs as the sternellum of the prothorax, to cover the tips of the epimera, and thus the coxal cavities become closed. In *sep* the epimera are variously short; in the extreme case they fail to reach the sternellum by almost the equivalent of the width of this structure, the coxal cavities becoming open (Plate 13,E,F). The shortening of the

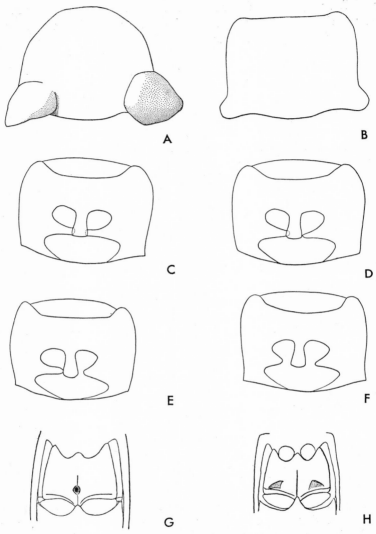

PLATE 13. Various mutants having the thorax affected. (A,B) Knobby prothorax (*knp*) pupa and adult, respectively. (C) Normal prothorax, ventral view. (D) Deflected epimera (*dep*). (E,F) Two examples of separated epimera (*sep*). (G) Umbilicus (*umb*). The stippled area represents a hemispherical depression on the median groove. (H) Scar (*sc*). The stippled areas in the metathorax are shallow sclerotized depressions. (G by permission of the Editor, *Canadian Journal of Genetics and Cytology*.)

epimera may not be symmetrical, more unsclerotized tissue showing to the left (or right) of the sternellum than on the other side. In the mildly expressed classifiable beetles the tips of both epimera almost reach the sternellum. Selection for stronger expression of this gene is being attempted. No linkage information (Sokoloff, 1965c).

(3) Deflected epimera (*dep*, Hoy and Sokoloff, 1964). In the normal beetle the epimera extend medially behind the coxae of the first pair of legs, and their tips lie under the sternellum (Plate 13,C). In *dep* the tip of one or both epimera are deflected, ending slightly above or behind the sternellum (Plate 13,D). Genetically it behaves as if incompletely recessive. It is not allelic with *sep*.

f. Genes Affecting the Metathorax

(1) Umbilicus (*umb*, Sokoloff, 1962). A recessive of variable expression and incomplete penetrance which appeared spontaneously, once in the pearl stock and again in a sample from a wild population during a determination of the genetic load in *T. confusum*. In the normal metathorax there is a shallow median groove which separates the two sternella. This groove is about three-fifths as long as the metasternum. In the *umb* mutant there appears a hemispherical depression of variable depth on the median groove (Plate 13,G). The *umb* phenotype can be readily identified in the adult, but in the pupa this area is hidden by the meso- and metathoracic appendages. It is doubtful that this character can be identified in the larva. No linkage is apparent between *p* and *umb* (Sokoloff, 1964).

(2) Scar (*sc*, originally described as engraved metasternum, *ems*, Sokoloff, 1962). Appeared spontaneously in the same wild population being tested for its genetic load, and in the same individuals exhibiting *umb* (Sokoloff, 1963e). As in *T. castaneum*, this condition results from a gene with good viability, incomplete penetrance, and variable expression. Characterized by an irregular, roughly triangular depression midway between the median groove and the pleural sclerites, and anterior to the transverse groove of the metasternum (Plate 13,H).

(3) Incomplete metathoracic projections (*imp*, Sokoloff and Hoy, 1964). Spontaneous in a stock bearing *p* and possibly *umb* and *sc*. The posterior metathoracic processes, which normally produce an inverted V to accommodate the median anterior process of the first apparent abdominal segment, is missing (Plate 7,D). In some individuals the median groove is deflected, usually to the right, and the metasternum may appear protuberant. Penetrance poor (Sokoloff, 1965c).

g. GENES AFFECTING ABDOMINAL STRUCTURES

Aside from the sex-linked *pas* gene, the following alter the phenotype of the abdominal sternites.

(1) Creased abdominal sternites (*cas*, Sokoloff, 1963). A recessive of good viability, incomplete penetrance, and variable expression, discovered in crosses attempting to establish linkage between *p* and *dpe*. Characterized by the presence of a diagonal groove on the lateral fourth of the abdominal sternite (Plate 12,C). In extreme expression all the abdominal segments have such a groove, although that in the apparent first and second abdominal segments is longer than in the remaining abdominal segments. In weaker expression of the gene only the apparent first abdominal segment may bear a shallow groove, usually hidden by the hind legs, and it may be restricted to only one side of the body. It is located 37.62 ± 4.82 units away from pearl, the order of the genes being *p–dpe–cas*. The *cas* gene is present in practically all of our stocks (Sokoloff, 1964f). Stanley (1965) has found *cas* in a sample derived from one of Dr. Thomas Park's laboratory strains designated as Chicago wild type.

(2) Sternites incomplete (*sti*, Sokoloff, 1963). A recessive of good viability but variable expression. Adjacent sternites of the abdominal segments fail to overlap, and in extreme expression an area almost half as wide as the typical abdominal segment may be unsclerotized (Plate 12,D). In less extreme expression small, bilateral, recessed, unsclerotized areas equidistant between the midline and the pleurite appear, and these may accumulate flour. Spontaneous in studies on genetic load of populations reconstituted from several inbred strains (Sokoloff, 1964h).

(3) Medial abdominal groove (*mag*, Sokoloff and Hoy, 1964). Two *mag* males and one female were found in the F_2 males derived from the *St* stock being tested to see whether a bleached condition of the elytral tips of a male was heritable. (It was not.) Autosomal recessive of variable expression and incomplete penetrance (only about 20% of the beetles will exhibit the character in the progeny of *mag* × *mag* matings). The mutant is characterized by the presence of a shallow medial depression in the last two abdominal segments visible when the light reflects at a certain angle from the sternites (Plate 12,E). In weakly expressed mutants only the penultimate segment will show a shallow depression. So far as is known this character cannot be observed in the pupa (Sokoloff, 1965c).

(4) Reduced juvenile urogomphi (*rju*, Hoy and Sokoloff, 1964).

Found among the progeny of b/+ sib matings. The appearance of this mutation is similar to that of *rju* in *T. castaneum:* a pair of small bristles appears, one on each side of the anal opening in both sexes (Sokoloff, 1965c).

h. Genes Affecting the Antennae

(1) Fused antennal segments-1 (*fas-1*, Sokoloff, 1962). A recessive of incomplete penetrance and variable expression (Sokoloff, 1962d). In a sample of the *fas-1* stock, about half of the beetles did not show antennal deformities. Those showing fusions of, or deformed, adjacent antennameres irrespective of right- and left-handedness were distributed as shown in Table 14. The mutant is illustrated in Plate 14,A,B. It is noteworthy that when a double fusion occurs in segments 7–8 and 9–10 on one antenna, the evidence for a two-segmented origin for that part of the antenna is a lateral incision for segments 7–8, and a medial incision for segments 9–10 (Plate 15,B, left). But if a segment intervenes, both fused masses are incised laterally (Plate 14,B, right).

(2) Fused antennal segments-2 (*fas-2*, Sokoloff, 1963). This spontaneous recessive mutant is very easy to identify even by examining beetles with the unaided eye, since, through complete fusion of segments 3–4 and 5–6 of the funicle, and secondarily of some segments of the club which is considered to consist of five segments (Hinton, 1948), the antenna becomes visibly shorter (Sokoloff, 1964h) (Plate 13,C,D). A sample of ten males and ten females from one of the established stocks

TABLE 14

Antennal Fusions in a Sample of *fas-1* in *Tribolium confusum**

Segments fused in both antennae		Male	Female
0	7–8	2	6
7–8	7–8	4	10
7–8	7–8, 9–10	4	2
7–8	9–10	1	0
0	9–10	1	0
9–10	9–10	0	1
7–8, 9–10	9–10	1	0
7–8	7–8, 9–11	0	1
		20	20

* Irrespective of left- and right-handedness.

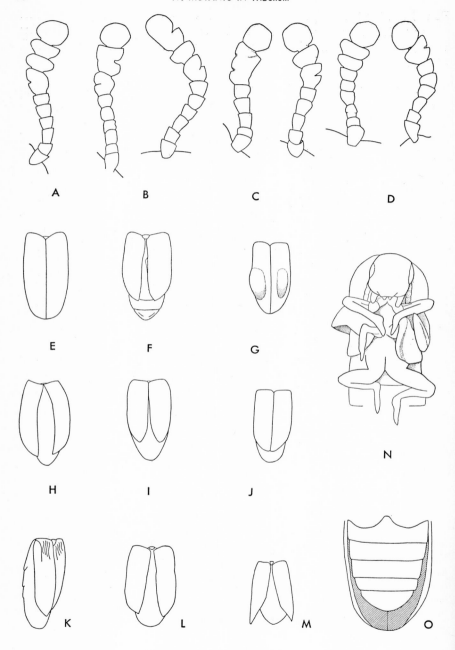

B. *Tribolium confusum* 99

Antennal Fusions in *fas-2* Male and Female *Tribolium confusum**

Segments fused in antennae		Male	Female
3–4, 5–6	3–4, 5–6	2	3
3–4, 5–6	3–4, 5–6, 9–10	0	3
3–4, 5–6, 9–10	3–4, 5–6, 9–10	0	1
3–4, 5–6, 8–9	3–4, 5–6, 8–9	8	2
3–4, 5–6, 8–9	3–4, 5–6, 8–11	0	1
		10	10

* Irrespective of left- or right-handedness.

gave the distribution of phenotypes given in Table 15. (However, in other stocks club segments 8–9 or 9–10 were rarely fused.)

i. Genes Affecting the Elytra

Aside from the sex-linked *aer*, *ma*, and *tet* genes which affect the elytra in various ways, the following autosomal mutants have been found:

(1) Split (*sp*, McDonald, 1959) is a recessive which in homozygous condition shortens and separates the elytra exposing the posterior third of the abdomen (Plate 14,F). Fertility of *sp* is lowered as indicated by a high percentage of infertile pair matings (about 37%) and viability is reduced by about 12% (McDonald, 1959a).

The *sp* and *b* genes appear to be in different linkage groups in some crosses (Sokoloff, 1964); in others the two genes appear to be linked, but loosely so (Sokoloff, 1965b). More data with other genes in this linkage group are desirable.

(2) Blistered elytra (*ble*, Sokoloff, 1961) appeared spontaneously in the Chicago wild-type strain. It is a recessive producing a huge blister

PLATE 14. Various mutants affecting the appendages in *Tribolium confusum*. (A,B) Right antenna and left antenna of two examples of fused antennal segments-1 (*fas-1*). Note orientation of fused blocks when they are adjacent and when a segment intervenes between fused blocks. (C,D) Two pairs of antennae derived from the fused antennal segments-2 (*fas-2*) strain. Note reduction in total number of antennameres. (E) Normal elytra. (F) Split (*sp*). (G) Rough (*ro*). (H) Raised split (*rsp*). I Disjoined (*dj*). (J) Short elytra (*sh*). (K,L) Semilateral and dorsal view of thumbed (*thu*). (M) Blade (*bld*). (N) Nude (*nd*) pupa. Note short elytra and membranous wings. (O) Elongated elytra (*ele*). The shaded area represents the ventral surface of the elytra. Note that the elytra exceed the length of the abdomen by nearly the length of the posterior abdominal segment.

on one or both elytra (Plate 9,R,S). This blister usually is located in the middle of the elytra, and in this respect differs from the *ble* mutant in *T. castaneum,* since in this species the blister occurs most frequently at the distal end of these appendages but less frequently it may be found at the proximal portion or on the lateral margins of the elytra. The blisters develop primarily at eclosion, although an occasional pupa may be found with blisters already formed. Viability of *ble* depends on the size of the blister and the degree of distortion of the elytra: if the blister is large the elytra may be lifted away from the abdomen with consequent loss of moisture and accumulation of flour (Sokoloff, 1961b).

(3) Rough (*ro*, McDonald, and Peer, 1961) appeared spontaneously in the McGill *b* strain. It is a recessive of variable expression. Many pupae appear normal while others bear a blister on one or both elytra. If pupae have blisters, the few surviving into the imago will also exhibit blisters on the elytra (Plate 14,G). Beetles appearing normal in the pupa eclose normally but after sclerotization will appear with their elytra warped or wrinkled. Selection for bilateral blisters increases the frequency of blistering. Selection for right or left blisters has failed to produce an asymmetrical change, which seems to indicate that the genes affect both sides of the body equally (McDonald *et al.,* 1963).

(4) Short elytra (*sh*, Ho, 1962). Spontaneous in an inbred line derived from a synthetic strain which had been propagated by brother-sister matings for more than twenty-three generations. It is a recessive with complete penetrance but somewhat variable expression. The elytra are short, exposing the posterior third of the abdomen (Plate 14,J). Short elytra differs from *sp* in that the posterior edges of the elytra meet at the midline. Preliminary crosses between *sh* and *sp* produced an F_1 with slightly shorter elytra than normal, but longer than in either mutant (Ho, 1962a). At this writing other tests of allelism or linkage tests have not been performed.

(5) Droopy elytra (*dre*, Sokoloff, 1961). Spontaneous in McGill *b*. Possibly a phenodeviant. The black stock repeatedly produces a small proportion of individuals (about 3–5%) with elytra divergent, sometimes starting at the scutellum (resemble Fig. H in Plate 14). Crosses of *dre* × *dre* yield a large number of normals, but with repeated selection the frequency of *dre* increases to over 50% (Sokoloff, 1961, unpublished).

(6) Warped elytra (*we*, Sokoloff, 1963). Spontaneous in *es*; *fas-1* × New York +/+ crosses. Possibly a phenodeviant, but a further check of the membranous wings of teneral adults is needed to determine whether the warped phenotype is produced by a blister in the hind

wings and a failure of the fluid from the blister to return to the body cavity before sclerotization of the elytra is accomplished (Sokoloff, 1963e).

(7) Elongated elytra (*ele*, Sokoloff, 1963). A recessive of variable expression which appeared spontaneously in experiments attempting to establish the map position of *aer*, *es*, and *lp*. A pair mating of normal sibs produced a number of beetles with elytra extending the equivalent of the whole posterior abdominal segment beyond the tip of the abdomen (Plate 14,O). The phenotype of *ele* may overlap that of normal beetles. Viability of strongly expressed *ele* appears to be reduced. It is probable that the extremely long elytra of *ele* females may prevent successful insemination by any male (Sokoloff, 1964h).

(8) Disjoined (*dj*, Dawson, 1962). A recessive appearing spontaneously in a suspended selection line, characterized by a divergence of the elytral tips much like that exhibited by *dve* in *T. castaneum:* the elytra may diverge starting at, but most often one-third to one-fourth of the elytral length behind, the scutellum, and the tips may gradually taper towards the sides. The posterior edge of the elytra may produce an attenuated sigmoid curve. The elytra may appear shorter by nearly the length of the exposed posterior abdominal segment (plate 14,I). Identifiable in the pupa by a clear reduction of the elytra and membranous wings and exposure of the tarsi of the hind legs, which in normal pupae do not extend beyond the elytral tips. No significant reduction in viability. Not linked with pearl, black, or ebony, and not allelic with *sp*, *sh*, or *thu*[s] (Dawson, personal communication, 1964).

(9) Thumbed (*thu*, *thu*[s], Dawson, 1963). The first allele was found in an inbred line. It is an autosomal recessive. The elytra appear shorter by almost one abdominal segment. The posterior edges of the elytra meet at the midline, and yet the distal portion of the elytra appears lifted off the abdomen giving these structures a vaulted appearance. The elytra bear an indentation located one-fifth to one-sixth of the elytra length behind the scutellum, as if the elytra had been pushed down at this point. Some *thu* beetles may exhibit a tiny split condition of the elytra. In most beetles, the membranous wings also appear shorter than normal and the distal ends fail to fold under the elytra. The allele *thu*[s] (originally reported as *rsp*, Dawson, 1964) was found in a subline isolated from a line selected for thirteen generations and in which selection had been suspended for one generation (and which also produced *dj* above). This allele of *thu* has elytra shortened and usually widely divergent (Plate 14,K,L). When tests of allelism were per-

formed a maternal effect appeared to be in operation since the phenotypes of the progeny tend to resemble that of the mother, and it
became evident that *thu* is partially dominant to *thu*^{s}. In spite of
selection against *thu*^{s} in the selected line the frequency of this gene has
remained at a frequency of 0.28–0.32. Dawson (1964, personal communication) believes that *thu*^{s} occurred within a linked complex of
genes which, when heterozygous, promotes fast development.

(10) Blade elytra (*bld*, Sokoloff and Hofer, 1964). In a single pair
mating possibly involving *em* (q.v.), eleven male and eight female pupae
were isolated because they appeared to be similar to pointed elytra in
T. castaneum. Seven pupae became imagoes but died without leaving
progeny. They had the following defects: (*a*) elytra divergent, sometimes starting at the scutellum, sometimes farther back, and somewhat
narrower but as long as the normal elytra (Plate 14,M). (*b*) The distal
fourth of the elytra is gradually reduced in width terminating in a sharp
point, or the tips may be narrow but not pointed and slightly curved
toward the midline, or the distal fourth may be almost normal but depressed toward the abdomen; or the tips of the elytra may be blistered
and pointed away from the abdomen. (*c*) The legs beyond the tibiofemoral joint may be missing, but this may have been due to a quinone
effect. Other deformities not attributable to quinones were present: the
epimera were shorter than normal, failing to extend under, and to fuse
with, the prosternum, or they were altogether wanting. In extreme expression the abnormality is increased to the point that the epimeron (E),
and the trochantin (T), and that part of the episternum separating E
and T are missing, so that the front coxae are open behind and may
be completely uncovered. The sternellum of the mesosternum may be
narrowly or widely separated from the anterior medial projection of the
metasternum in a manner similar to the condition produced by *ims* in
T. castaneum (Sokoloff, 1965c).

(11) Nude (*nd*, Hoy and Sokoloff, 1964). Spontaneous in crosses attempting to establish the mode of inheritance of *em*. One male and two
female pupae were found with distal ends of the elytra folded under
as in the *T. castaneum te*. The male imago had one elytron bent, and
the other appeared folded under, but it had a huge blister. The F_1 were
normal; the F_2 exhibited a wide range of elytral abnormalities: in some
pupae the elytral and membranous wing buds were greatly reduced,
and the elytra were tucked between the prothorax and the first two pairs
of legs (Plate 14,N). The third pair of legs was exposed. These extremely
deformed pupae failed to eclose. Among the F_2 imagoes were large num-

bers of beetles bearing fairly normal-sized elytra and membranous wings but variously split. At this writing it is known that the character is heritable, but the mode of inheritance is to be determined (Sokoloff, 1965c).

j. Genes Affecting the Legs

(1) Deformed legs (*dl*, Sokoloff, 1962). Probably a phenodeviant, found in crosses between e_2 and *p*. Usually the deformity is confined to the femur and frequently only one pair or only one member of the pair of legs is affected (Plate 15,B). The femur may be short and pear-shaped, or short and bent near the coxofemoral joint. In some cases the tibia also is affected, becoming variously curved, and the tarsal segments may be reduced so the onychium is attached directly to the femur (Sokoloff, 1962d).

(2) Bent tibia (*btt*, Sokoloff, 1962). Spontaneous in crosses determining the map position of l_1. Possibly a phenodeviant. The proximal end of the hind tibiae, in strong expression of the character, has a double sharp angle resulting in an apparently shorter leg. In weaker expression the tibia acquires a gentle curve (Sokoloff, 1964h).

(3) Bent femur (*btf*, Hoy and Sokoloff, 1964). Found while selecting for a pure stock of *es tet*. About thirty individuals of both sexes were found with very strong expression of bent tibia (*btt*). The *btf* differs from *btt* in having, in addition, a deformed femur. The deformed femur can occur in any pair of legs, but it is abnormal only in those legs exhibiting a strong expression of the bent tibia phenotype. The deformity consists in having the proximal third of the femur somewhat narrower than the rest, and at this point the femur is sharply bent, producing an angle of approximately 135° between the two parts of the femur. Tests of allelism with *btt* have not been performed (Sokoloff, 1965c).

k. Genes Affecting Larval or Pupal Urogomphi

No mutants comparable to *eu* or *u* in *T. castaneum* have been found so far in *T. confusum*.

l. Genes Affecting the Genitalia

Emasculated (*em*, Sokoloff and Hoy, 1964). Found in several linkage-test crosses: In the F_2 of $+/e$; $+/umb$; $+/p$ mated *inter se*, five males were found which, on squeezing, appeared to have no aedeagi. A number of males with no apparent aedeagi was found in some backcrosses of F_1 females of the cross *p/p; umb/umb; ems/ems; cas/cas; sti/sti* (*imp?*)

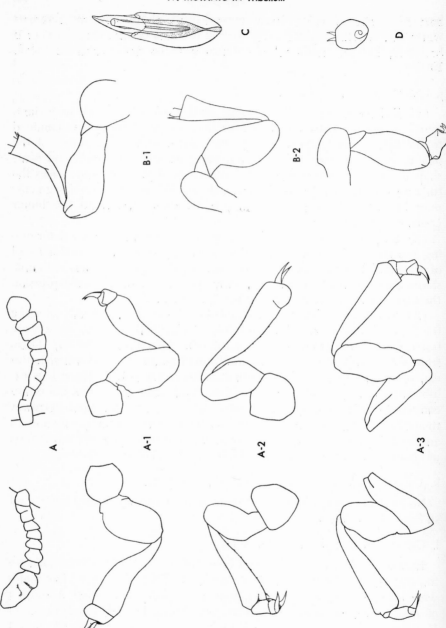

males which also had short epimera (*sep*, q.v.) × normal females back
to the P$_1$; and finally, one male was found in the F$_2$ of crosses between
e and *b*. The *em* males, superficially examined, resemble normal males
in the possession of normal basal pits with associated hairs on the femurs
of the first and second pairs of legs (the latter being less pronounced),
but no aedeagi are extrudable when gentle pressure is applied to the
abdomen. Dissection of these beetles reveals normal internal reproductive organs (testes and associated glands and structures) connected to
a ball-like sclerotized structure (Plate 15,D) sometimes connected to a
small projection which may be an incipient aedeagus, located entirely
within apparent abdominal segment III, or over the suture separating
apparent segments III and IV. Matings of these beetles, which survive
for a long time, are apparently sterile because they are incapable of
transferring their sperm. It is possible that some *em/em* males may have
short aedeagi, but more work is needed to clarify this point. The results
of various crosses seem to indicate that *em* is an autosomal recessive
sex-limited gene of variable expression. Preliminary crosses fail to establish linkage between *em*, *p*, and *rus* (Sokoloff, 1965c).

m. GENES AFFECTING THE BIOCHEMISTRY OF THE FLUID IN THE STINK
GLANDS

Melanotic stink glands (*msg*, Ho and Sokoloff, 1958, 1963). Spontaneous. Discovered (by Ho) in various laboratory strains, and
subsequently (by Sokoloff) in the Berkeley synthetic strain in crosses
connected with an investigation of the genetic load in *T. confusum*. Autosomal recessive of good penetrance and viability and variable expressivity. Typically the prothoracic stink glands (PSG) are more frequently
affected than the abdominal stink glands (ASG). The PSG appear as
dark spots at both anterior projections of the prothorax, the spots being
rounded, elliptical, or triangular. The material in the reservoirs turns
black in 2 or 3 weeks after eclosion, and it may form a single compact
mass or it may be broken into several components of different size (Plate
12,F). The ASG are less reliable for classification of *msg* since neither,
only one, or both may be pigmented, and usually they acquire pigmenta-

PLATE 15. Various mutants in *Tribolium confusum* (A) The antennae of stilted
legs (*stl*). A-1 to A-3 are the legs from the same individual. Note reduction in the
size of the femur and in the number of tarsomeres. (B) Three legs of deformed
legs (*dl*). (C) Normal aedeagus. After El Kifl (1953). (D) The aedeagus in emasculated (*em*). (C by permission of the Editor, *Bulletin de la Société Entomologique
d' Egypte*.)

tion later than the PSG. Squeezing of the abdomen may empty the contents of the PSG and then they cannot be seen. The fluid so squeezed out has dark particles which feel crystalline when touched with forceps. Dissection of the prothorax indicates that the material in the PSG is also crystalline, forming a lump which may fill the whole reservoir of the gland, or if smaller it is surrounded by a colorless fluid (Sokoloff, 1963e).

Chemical analysis of normal beetles indicates that the secretions contained in the reservoirs of the odoriferous glands consist of the same quinones as those found in *Tribolium castaneum* (see above) except that 2-methoxy-1,4-benzoquinone was absent (Hackman *et al.*, 1948; Roth and Eisner, 1962). Analysis of *msg* indicates that young beetles may have only about one-twentieth the amount of quinones of normal beetles, but in older beetles no quinones were detected (Engelhardt *et al.*, 1965).

When flour beetles are placed in a confined environment with fresh food, they alter the flour through their feeding and metabolic activities (including the discharge of quinones) giving the flour a pinkish color, an offensive odor, and a disgusting taste. The flour is said to be "conditioned." This conditioning has different effects on fecundity and other biological attributes of *T. castaneum* and *T. confusum* (Park, 1934b, 1935, 1936; Park and Woollcott, 1937; Prus, 1961; Sonleitner, 1961).

The role of the odoriferous secretion in *Tribolium* has been under speculation. The secretions may have played a defensive role when *T. castaneum* and *T. confusum* occupied a different habitat, but as Roth (1943) has emphasized, at present *Tribolium* beetles encounter few predators in the flour they infest, and quinones (save in direct contact) appear to be ineffective in warding off mites which may be the beetles' chief predators in their present habitat. Van Wyk *et al.* (1959) found that *T. confusum* is generally attracted to flour containing storage fungi or bacteria isolated from the beetles themselves, but as the beetle population increases "the population of storage fungi decreases almost to the vanishing point, presumably because the quinones, secreted by the beetles, are toxic to the fungi." These authors believe that one of the functions of the malodorous secretion of flour beetles is to keep the food material relatively free of microorganisms. During our studies on *msg* it has been found that in crowded cultures of the mutant the medium becomes moldy in a relatively short period, whereas the medium in crowded normal cultures remains free of mold, lending support to the suggestion by Van Wyk *et al.* (1959), that the role of quinones is a bacterio- and fungistatic one, enabling flour beetles to grow in a medium

elatively free of microorganisms which, if allowed to grow without re-
traint, would compete directly with, or make the substrate unsuitable
or, the insect.

The *msg* gene has been found linked with, and about 42 recombination
nits away from, *b* (Sokoloff, 1964f).

. GENES HAVING PLEIOTROPIC EFFECTS

Aside from the sex-linked genes, *St* (which has lethal effects in homo-
nd hemizygotes, produces sterility in the surviving *St* males, and a tem-
orary light colored stripe in the middle of the elytra of females
genetically *St/+*) and *aer* (which reduces the length of the elytra, and
he number of antennameres and tarsomeres), only one autosomal re-
essive gene is known to produce multiple changes of the body:

Stilted legs (*stl*, Sokoloff, 1962) appeared spontaneously in ebony-2
crosses. The main noticeable effect is on the legs; the femurs become
hort, the tibiae long and slender, and the tarsi variously fused (Plate
5,A). Segments of the antennae may exhibit fusions of the terminal

TABLE 16
Antennal Fusions in Stilted Legs (*stl*) in *Tribolium confusum*

Male		Female	
Right	Left	Right	Left
0	0	7–8, 9–10	4–5
7–8	9–10	7–8	9–10
0	7–8	7–8, 9–10	7–8, 10–11
9–11	10–11	7–8, 9–10	7–8, 9–10
10–11	10–11	7–8, 9–10	7–8
9–10	0	7–8	7–8, 10–11
4–5, 7–8	10–11	0	7–8, 9–10
9–11	9–11	4–5	9–10
4–5, 9–10	9–10	4–5, 9–10	9–10
2–3, 4–5, 10–11	2–3, 4–5, 9–11	10–11	10–11
0	3–4, 10–11	7–8, 10–11	7–8, 10–11
5–6, 9–11	5–6, 9–11	9–11	10–11
7–8, 10–11	0	5–6, 9–10	5–6, 7–8
7–8, 9–10	7–8, 9–10	7–8, 9–10	7–8, 9–10
0	0	0	9–10
		7–8, 9–10	7–8, 9–10
		5–6, 9–10	7–8, 9–10
		7–8, 9–11	5–6, 9–11
		7–8, 9–10	7–8, 9–10
		0	5–7, 9–10

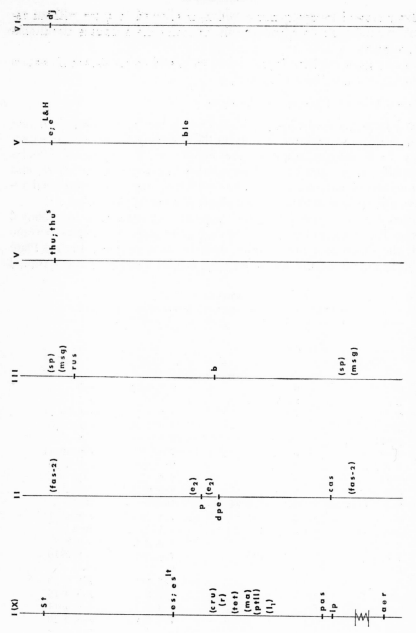

FIG. 2. Linkage maps in *Tribolium confusum*.

segments of the club and occasionally of more proximal segments (Table 16), and segments of the maxillary and labial palps also may be variously fused. The mutant is identifiable in the pupae. Imagoes may die at eclosion through failure to disengage themselves from the pupal skin or by being unable to enter the flour, lying on their backs after eclosion. Taken out of the flour, the beetles have difficulty in maintaining an upright position, and judging by the number of sterile crosses, the males may have difficulty in achieving a successful grasp of the female while attempting copulation. The *stl* gene is a recessive of good penetrance, variable expression in regard to the fusion of the antennal and tarsal segments, and reduced viability (Sokoloff, 1962d).

3. Linkage Maps in *Tribolium confusum*

The maps given in Fig. 2 are based on the accompanying linkage data, some of which have been published (Sokoloff, 1964f).

Linkage Relationships in *Tribolium confusum*

I	II	III	V
$es–es^{lt} = 0$	$p–dpe = 5$	$b–rus = 41$	$e–ble = 30–40$
$es–St = 38$	$p–cas = 38$	$b–msg = 42$	
$es–lp = 47$	$dpe–cas = 30$	$b–sp = 42–44*$	
$pas–lp = 3$	$p–e_s = 2.5$		VI
$lp–aer = 24$	$p–fas-2 = 42–44$	IV	$dj†$
$es–tet = 53$		$thu, thu^s†$	
$r–lp = 48$			
$es–l_t–40$			

* Some crosses give these recombination values, but others give values nearer 50%. Three-point crosses have not been carried out.

† Dawson (personal communication) has determined that *thu* and *dj* are not linked to *b*, *p*, and *e*. They are here, therefore, assigned as markers of linkage groups IV and VI, respectively.

In addition, the following have been tested and proved to segregate independently: *b* from *lod*, *fas-2*; *e* from *cas*, *ele*, *fas-1*, *fas-2*, *lod*, *p*, *rsy*, *sti*, and *stl*; *fas-2* from *ele* and *lod*; *rus* from *ble*; *p* from *umb*.

C. *Tribolium anaphe*

Only one mutant is known in this species.

Sternites incomplete (*sti*, Sokoloff, 1964). This occurred spontaneously in descendants of a sample kindly provided by Mr. C. E. Dyte, Pest Infestation Laboratory, Slough, Bucks., England. It is an autosomal recessive of good penetrance but variable expressivity, resembling the *sti* mutant in *Tribolium castaneum* (Plate 12,D) (Sokoloff, 1964h).

D. *Tribolium destructor*

Two mutations have appeared spontaneously in a stock derived from one maintained at the Pest Infestation Laboratory, Slough, Bucks., England. Both are autosomal recessives:

(1) Creased abdominal sternites (*cas*, Sokoloff, 1964). This incompletely penetrant, variously expressed gene produces grooves on the abdominal sternites, mostly on the apparent first segment. Phenotypically it appears identical to *cas* in *T. castaneum*, and *T. confusum* (Plates 8,B and 12,C) (Sokoloff, 1964h).

(2) Bent tibia (*btt*, Sokoloff, 1964). The legs of this mutant appear shorter because of a double bend in the tibiae. As in *T. castaneum* (Plate 7,L) the effect is more frequently observed on the hind legs but other legs may exhibit the abnormality, indicating that the gene has poor penetrance (Sokoloff, 1964h).

E. *Tribolium madens*

Four autosomal genes are known for this species. They were found in descendants of a sample obtained from the Pest Infestation Laboratory, Slough, Bucks., England.

(1) Split (*spl*, Sokoloff, 1963) is a recessive of variable expression, the extremely malformed beetles having the elytra divergent beginning at the scutellum, less extreme phenotypes having only a feeble split condition at the posterior end of the elytra. It resembles the *spl* mutant in *T. castaneum* (Plate 9,M) (Sokoloff, 1964h).

(2) Bent tibia (*btt*, Sokoloff, 1964), was found in crosses attempting to establish the mode of inheritance of *spl*. A recessive gene of poor penetrance and variable expression: in badly deformed mutants the hind tibiae exhibit a double sharp (about 90°) angle resulting in a much shorter hind leg. In weakly modified mutants the tibiae are curved. Resembles *btt* in *T. castaneum* (Plate 7,L) (Sokoloff, 1964h).

(3) Creased abdominal sternites (*cas*, Sokoloff, 1963). A recessive of variable expression and incomplete penetrance resembling *cas* in *T. castaneum*, *T. confusum*, and *L. oryzae* (Plates 8,B, 12C, 17D) (Sokoloff, 1964h).

(4) Fused antennal segments-1 (*fas-1*, Sokoloff, 1963). A recessive of variable expression. Club segments 9–10 are fused, in one or both antennae. Funicular segments seem not to be affected (Sokoloff, 1964h).

V. Genetic Studies in Other Coleoptera

It is highly probable that the scarcity of information on the basic genetics of beetles is largely the result of an improper choice, on the part of the investigator, of the organism to be investigated. In many cases beetles fail to reproduce in the laboratory. In those that do, technical difficulties in maintaining them under laboratory conditions (for example, Coccinellidae) or in obtaining virgins (for example, Bruchidae) have been encountered. In still others (for example, the tenebrionid *Tenebrio molitor*) the developmental cycle was forbiddingly long to make rapid progress. Despite these drawbacks the literature, particularly of the population genetics of coccinellid species, is extensive. Most of these studies, notably those of Dobzhansky (1924), Tan and Li (1932–1933, 1934), Timofeeff-Ressovsky (1940), and Komai and co-workers, have been concerned with analyses of phenotypic variation in relation to geographic distribution and ecological factors such as climate. The literature on this subject has been reviewed by Shull (1943), Komai *et al.* (1950), and Komai (1956). Here those studies concerned primarily with the determination of the mode of inheritance of specific characters will be cited. The beetles in this section have been grouped according to the taxonomic classification in Brues *et al.* (1954).

A. CUCUJIDAE

In a stock list submitted to the *Tribolium Information Bulletin* by Dyte (1963) for the Pest Infestation Laboratory (PIL), the following mutants have been observed in species of this family of beetles.

1. *Cryptolestes pusillus*

A black form has been isolated from a stock derived from Trinidad. Lefkovitch (1963) reports that this mutant, crossed with wild type, so far has given 133 brown and 55 black individuals in the F_2. This evidence does not disagree with the hypothesis that black is controlled by a single, fully penetrant, autosomal factor recessive to the normal brown color.

2. *Cryptolestes turcicus*

Dyte (in stocks kept at PIL) has found a red (*r*) eye mutation in this cucujid. Preliminary crosses suggest this is a sex-linked mutation, and apparently the first mutation of this type outside of the Tenebrionidae (Dyte *et al.*, 1965). Shaw (1965) has found that the *r* gene exhibits pleiotropy: the normally black or dark brown Malpighian tubules become pale or transparent.

In a wild-type sample obtained from PIL and cursorily examined, the following abnormal beetles have been found:

(*a*) Runty (*rty*, Sokoloff and Hoy, 1964). Normally males of this species are smaller than females. The abnormality cited herein is restricted to the females, which are reduced to about half the size of normal females. No males have been observed to be reduced to that extent. The fact that many females exhibit the phenotype indicates the condition is heritable, but the mode of inheritance has not been worked out (Sokoloff, 1965c).

(*b*) Crooked antennae (*cka*, Hoy and Sokoloff, 1964). Spontaneous in material from stocks maintained at PIL. Autosomal recessive of variable expression and incomplete penetrance. Usually segments 4 and 5 of one or both antennae become fused into a sausage-shaped segment of about the same length as the two separate segments (Plate 9,B). Sometimes more distal segments may be affected in the same individual; occasional individuals may exhibit a bifurcated antenna beyond the fused segments, but these may be teratologies (Sokoloff, 1965c).

B. SILVANIDAE

1. *Ahasverus advena*

Dyte (1963) reports a black body color mutant. The mode of inheritance of this mutation has not been determined.

2. *Oryzaephilus surinamensis*

Dyte (1963) reports a pearl-like mutant found in a sample derived from Australia. It is an autosomal recessive (Dyte *et al.*, 1965). Shaw (1965) reports the Malpighian tubules in *p* beetles are pale or transparent, while in normal beetles the tubules are black or dark brown.

C. DERMESTIDAE

1. *Dermestes maculatus*

(*a*) Philip (1940) found an autosomal recessive gene, "white eye." This mutant was said to be comparable to pearl in *T. castaneum.* Dyte *et al.* (1965) found a "pearl" (*p*) mutation possibly allelic with the now nonexistent mutation white eye. Its genetics has not been worked out, but Shaw (1965) states that this mutation has pleiotropic effects: the black or dark brown Malpighian tubules characteristic of normal beetles become pale or transparent in the *p* adults.

(*b*) Rufous (*ru*) is a recently discovered eye color mutation the genetics of which has not been worked out. The normally black eye becomes red-brown (Dyte *et al.*, 1965).

(*c*) Fuscous (*fu*). The normally black elytra are modified to a deep brown. Preliminary crosses suggest semidominance (Dyte *et al.*, 1965).

(*d*) Second sex pit (*ssp*). In addition to the sex pit normally present in males on the fourth apparent sternite, a sex pit appears on the third apparent sternite. May be a recurrence of a similar mutation described and found by Philip (1940) to be an autosomal recessive with sex-limited expression (Dyte *et al.*, 1965).

(*e*) Light wing (*l*). Costa, subcosta, radius, and sector and enclosed cells are unpigmented (Dyte *et al.*, 1965). May be a recurrence of a similar autosomal recessive mutation described by Philip (1940).

(*f*) Short elytra (*sh*). The genetics has not been worked out, but the elytra in this mutant are short, exposing the last abdominal segment (Dyte *et al.*, 1965).

2. *Trogoderma granarium*

Reynolds and Sylvester (1961) reported on the autosomal recessive pearl in this dermestid. Dyte and Blackman (1961) found that the larval ocelli and the frontal ocelli and compound eye in the adult of this species lack pigment.

D. NITIDULIDAE

Carpophilus dimidiatus

A pearl mutant has been reported by Dyte (1963) in a stock-list submitted by him for the Pest Infestation Laboratory, Slough, Bucks, England. Dyte and Blackman (1961) note that the ocelli of the larva and the compound eye in the adult were pigmentless and pearl-like as in

T. castaneum pearl. Amos and Scott (1965) give data to show that this condition is governed by a single, autosomal recessive gene. Shaw (1965) has found that *p* beetles have pale or transparent Malpighian tubules while in the normal they are black or dark brown.

E. BOSTRICHIDAE

Rhyzopertha dominica

A black (*b*) mutant has been isolated, but its genetics has not been worked out (Dyte *et al.*, 1965).

F. COCCINELLIDAE

1. The Genus *Hippodamia*

Many species of ladybird beetles have conspicuous elytral markings. Shull (1943) has shown that in *Hippodamia sinuata*, the spotted (*s*) and spotless (*S*) condition results from the difference in the effect of a single gene. *S* is not quite dominant; though a few heterozygotes are strictly spotless, most of them have reduced spotting. There is little or no overlapping of the phenotype of the heterozygote and that of the spotted homozygote. Studies within the species *Hippodamia quinquesignata* and in crosses between it and *H. convergens* led to the discovery of four principal pairs of genes affecting elytral pattern. *S* is dominant in *H. convergens* and is as effective in eliminating the spotted pattern in *H. quinquesignata*. The band across the anterior ends of the elytra was found to be recessive (*q*) to *Q*, the bandless (two separate spots) condition. The presence of a lateral spot behind this band was dominant, *T*, over the absence of this spot (*t*). The oblique postmedian band of *H. quinquesignata* was regarded as recessive to the separation of the spots whose fusion makes this band. There were some irregularities in the inheritance of three of these characters (*Q*, *T*, *F*) which required the assumption of modifying or accessory genes: in each of these three the dominance postulated was occasionally lacking or reversed (Shull, 1945). In *H. convergens* experiments it was estimated that there were three or four modifying genes which restored a partially spotted condition in genetically spotless beetles. Populations derived from different geographic areas differed in the frequency of modifiers (Shull, 1944). The fortunate circumstance that in *Hippodamia* species hybrids can be obtained led Shull (1946) to a study of genes affecting the shape of male genitalia. He concluded, from crosses involving *H. quinquesignata* and *H. convergens*, that "the siphonic flaps and the aedeagal keel may e differentiated in the two species, each by two pairs of genes, the width of the sipho by five pairs, with a distinct possibility that the number could be three pairs for each of them."

. The Genera *Coelophora* and *Cheilomenes*

Dobzhansky (1933) analyzed the data of Timberlake (1922) on spotting and color pattern in *Coelophora inaequalis* and *Cheilomenes exmaculata*. Dobzhansky concluded that in the first species a series of three multiple alleles is involved: the nine-spotted condition is dominant o normal and the latter is dominant to black. In the second species Mendelian inheritance was also evident.

. The Genus *Harmonia*

Tan and Li (1934) elucidated the mode of inheritance of elytral patterns in *Harmonia axyridis*. The plain yellow pattern found in the *succinea* type was found to be determined by a single gene recessive to ny form with black-margined patterns. Three types of patterns among the black-margined beetles, known as *conspicua* (*C*), *spectabilis* (*S*), nd *aulica* (*A*) differed by a single Mendelian factor with *C* epistatic over *S* and *S* over *A* in a series. The authors concluded that the genetic ormula for *succinea*, *frigida*, and *19-signata* is *aa ss cc;* for *aulica AA s cc;* for *spectabilis aa SS cc* or *AA SS cc* and for *conspicua aa ss CC, AA ss CC, aa SS CC,* or *AA SS CC.*

The work of Tan and Li (1932–1933), Tan and Li (1934), and of Hosino (1940) has led to the recognition of eight patterns in *Harmonia axyridis* which constitute a multiple allelic series. Dominance between these alleles is said to be complete, partial, or lacking (Shull, 1943).

. The Genus *Adalia*

Lus, (1928), who studied *Adalia bipunctata,* concluded that eight different elytral phenotypes are under the control of genes at the same locus, but Timofeeff-Ressovsky (1940) feels that sufficient evidence for his conclusion is not available. Lus also gives data on three alleles affecting the elytra of *A. decempunctata.*

. The Genus *Epilachna*

Other than single genes or multiple alleles affecting color or spotting patterns, only one gene is known in the Coccinellidae to produce abnormal body traits: Timofeeff-Ressovsky (1935) found "Divergens," an autosomal dominant having recessive lethal effects in *Epilachna chrysomelina.* In this mutant the elytra diverged considerably, they were short and drooped at the sides, and viability of the heterozygote was greatly reduced.

G. TENEBRIONIDAE

1. *Tenebrio molitor*

A group of European investigators studied heritable characters in the meal worm. Arendsen Hein (1920, 1924a,b) who initiated the work, found a number of genes affecting larval body color, certain appendages, the head and the eye, and eye color. After Arendsen Hein's death the work was continued by Ferwerda (1928) and extended, chiefly along physiological channels, by Schuurman (1937). The following summarizes briefly the main findings of these investigators.

a. BODY COLOR

Arendsen Hein found larvae of three color types: orange, yellow brown, and umber brown. They differed from each other by one gene, the sequence of dominance being orange → umber brown → yellow brown. Ferwerda noted a reversal in dominance during metamorphosis; in the cross umber brown with orange the larva looks like the orange but the imago like the umber-brown type.

b. EYE COLOR

The normal color of the ommatidia, as in all tenebrionids, is black. Three recessive eye color mutants were found: the autosomal red (*f*), and flesh-colored (*g*), and the sex-linked yellow (*h*). The combined efforts of all three investigators have determined that eye color is determined by a series of interacting genes, the genotype producing the various phenotypes as shown in the tabulation:

Genotype		Phenotype
Female	Male	
FF GG HH	*FF GG Ho*	Black
ff GG HH	*ff GG Ho*	Red
FF GG hh	*FF GG ho*	Yellow
ff GG hh	*ff GG ho*	Yellow
FF gg HH	*FF gg Ho*	Flesh-colored
ff gg HH	*ff gg Ho*	Flesh-colored
FF gg hh	*FF gg ho*	Flesh-colored
ff gg hh	*ff gg ho*	Flesh-colored

From the phenotypic effects it is evident that *h* is dominant over *f*, and *g* is epistatic over *f* and *h*.

Schuurman (1937) found that *g* produced somatic mutations, a beetle with one eye black and the other flesh-colored, having been found and tested. In addition, a mosaic-eyed condition (with black spots scattered throughout the red eye) was discovered. Two hypotheses were advanced: either the black-spotting results from a mutable gene during development of the beetle's eye resulting in a mosaic eye, or it has resulted from a translocation of a piece of chromosome carrying the *F* gene to the chromosome bearing the *f* gene. This translocation may have been present in some cells of the body (including the sex cells) of the parents which produced F_2 progeny with mosaic eyes.

c. Head and Eye Abnormality

Ferwerda determined that the abnormality (characterized by a V-shaped groove and a reduced eye) found by Arendsen Hein was controlled by a dominant gene *B*, and that *B* and *g*, which are linked, have a lethal effect. Schuurman, however, was able to maintain *BB gg* beetles for at least two generations, so that this genotype is not always perfectly lethal. From repulsion backcrosses involving these two mutants in the two sexes Schuurman obtained 8.66 and 6.91% crossovers for females and males, respectively, indicating that crossing-over was not much different in the two sexes.

d. Appendage Abnormalities

Aside from abnormalities in the antennae and/or tarsi which were manifested by an increase or decrease from the normal number of antennameres or tarsomeres, but proved to be nonheritable variations, Arendsen Hein (1924a) found that the following were indeed mutations:

(1) Antennae compressed (*At c*) is characterized by the general compression of the body and of the appendages. The pleiotropic effects of this gene are described as follows:

(*a*) Antennae, tarsi, and prothorax are compressed and shortened in the axial line; these organs have lost their slender appearance; the joints are broadened; the front horns are rounded; the scutellum is often pseudosemicircular.

(*b*) The reductions either in antennae or tarsi are secondary; their number may be normal while these organs still show the characteristics specific to this strain.

(*c*) Complete fusions of two or more members give antennae in which ten, nine, eight, or even seven joints may be counted. A fusion of the fourth and fifth joints occurs often, mostly accompanied with a partial or whole fusion of the more peripheral members.

(*d*) "Reductions in the tarsi with one joint occur, but are far from general."

(2) Antennae and tarsi reduced with one joint (*At 10/10*) is characterized by a reduction of one segment in both antennae and all tarsi, resulting in beetles showing the composition of 10/10 segments in the antennae, and 4/4, 4/4, 3/3 in the tarsi. Penetrance is incomplete (only 88% *At 10/10* exhibited defects), and the abnormalities of these beetles are by no means uniform. The antennae and tarsi keep their usual slenderness. Reductions in the length of the prothorax and a rounding of its front horns may occur, but not as often as in *At c*. Some individuals exhibit reduced antennae and reductions may occur in one or more tarsi; the antennae may be reduced but the tarsi normal; the antennae may be normal but the tarsi reduced by one segment; or the antennae may be normal and one or more (but not all) tarsi reduced by one segment.

(3) Tarsi abnormal (*Ta*) is a gene of complete penetrance but variable expression; "not only are the conditions in two beetles scarcely ever exactly alike, but the same may also be stated of the six tarsi in one individual, which, as a rule, all show a different aspect. The conditions observed may be compared to those obtained with a telescope tube in which five members may be pushed in, or drawn out in different combinations, and by which *one*, or *two*, or *three* or *four* or even all *five* joints may entirely disappear, in which last case the terminal claws originate from the tibia."

Crosses of *At 10/10* × *Ta* gave a clear dihybrid segregation indicating they were not linked, but *At c* × *Ta* in two crosses gave an excess of *At c–Ta* over the two other possible combinations. It was not clear whether the two genes were linked or the deviation in the F_2 due to chance, and the experiment was not repeated.

During his intensive investigations, Arendsen Hein also noted certain abnormalities which in today's terms might be termed as phenodeviants. One pertained to an increase in tarsomeres in one or more tarsi to five or six segments. The incidence of this abnormality in the F_2 of crosses where one P_1 was abnormal was of the order of nearly 2%. The other abnormality was called *Px* (after the resemblance of the distal antennal segment to a Pixavon bottle, broad at the base and with a tapering neck, seen in European hairdresser shops).

One *Px* beetle crossed with a normal gave one male among the twenty

F_1 progeny with this condition and two males among the sixty F_2 progeny, but the F_3 were all normal. In a cross of two beetles showing a bilateral, well-expressed abnormality, forty-four F_1 (out of a total of fifty or 88%) had at least one antenna abnormal; the remaining 12% were normal. The abnormal F_1 were mated *inter se*, and of the thirty-three F_2 individuals obtained, 33% were normal and 67% had abnormal antennae. Arendsen Hein believed that this condition is a simple non-heritable modification, but pointed out that by continued selection and inbreeding it might be possible to obtain a strain "pure for the presence or absence of a factor underlying this anomaly, if such a factor exists."

Presumably the mutants found by Arendsen Hein, Ferwerda, and Schuurman are no longer in existence. To the writer's knowledge, the only mutation extant in this species is that found in populations of *Tenebrio molitor* at the Pest Infestation Laboratory (Dyte, 1963). This mutation resembles pearl and is probably a recurrence of "flesh-colored." Cursory examination of a population of this species by the writer yielded a few "droopy elytra" -like and "bent tibia" -like beetles, but they failed to reproduce in our medium. In view of the fact that *Tenebrio* is a tenebrionid, it would be useful to resume studies on the basic genetics of this organism to enlarge our knowledge of the comparative genetics of the Tenebrionidae.

Perhaps because of the long developmental cycle of *Tenebrio molitor* (9 months elapsing before a generation is completed at room temperature), little work has been done on the genetics of this species since the work of Arendsen Hein, Ferwerda, and Schuurman. Most of the recent literature utilizing this species is in the realm of physiological, nutritional, and other studies. A notable exception is a recent contribution by Leclercq (1963). This investigator started artificial selection experiments for small (F) and large (G) body weights in 1946, discarding pupae weighing more than 160 mg and later more than 130 mg in the F strain and all pupae weighing less than 160 mg in the G strain. From 1952 on, after twelve consecutive generations of selection for both strains, little selection was needed. It became evident that selection for body weight also had resulted in an automatic selection for shorter or longer durations of larval development. An assay of the two strains indicated that the two strains differed in a number of attributes: The F strain differs from G in producing

. . . advanced larvae, pupae and adults of bigger size and heavier weight. Its larvae reach the pupal stage more rapidly (3 generations per annum being possible instead of 2). Its pre-pupal period (larval age with lower growth rate and reduced nutritional requirements) is shorter and less variable. Such differences in the developmental rates are correlated with discrepancies in the growth of the phosphorus content. Very young F-larvae are more affected by cold. F-larvae are more favored when the food supplied is more truly optimal either by addition of yeast or when each larva is reared individually with superabundant flour. Nevertheless, the same larvae are less affected when fed with cereal flours of poor quality or with artificial diets, especially with carnitine deficient mixtures. On the other hand, they may well need more potassium. Adults of the F-strain are much more prolific and lay proportionately fewer fertile eggs. But the same adults are less affected in their fecundity if their larvae suffered from malnutrition, and they respond more positively when allowed to drink water.

Clearly, several of the revealed characters are correlated and controlled by the same genetic features. Nothing is known of the genetics of the two strains and no crossbreeding was tried. Nevertheless, one cannot imagine that such a complex of statistical differences compatible with a still great individual variability should be controlled by a single allele gene system. Thus, we must conclude that, by selecting for weight only, one has induced a process of complex differentiation involving a number of biochemical characters. It must be emphasized that no morphological or structural difference was detected.

Leclercq believes that these recorded differences have adaptive and evolutionary significance.

Already in standard incubator conditions a small initial population of the F-strain will take advantage of its faster developmental rate to produce a second generation, then a third one . . . each being more numerous than the previous one. It will be so not only because adults will appear more rapidly but also because they will appear within a shorter period (±7 weeks) the chances of mating being greater. On the other hand, a small initial population of the G-strain will encounter difficulties in producing further increasing populations owing to its longer larval life and to the fact that its adults will hatch within a much more extensive period (more than 12 weeks), the chances of mating being reduced. But this weakness will be compensated by the much greater fecundity (3 or even 5 times greater). In Nature and in warehouses, the factor "chances of mating," is likely to have a still more selective value because the adults wander far when they have hatched, also because the life-cycle of the species is not phased genotypically with the seasons. Here again speed of growth versus fecundity provides compensatory characters. Perhaps the G-strain will suffer from its lower resistance to adverse nutritional conditions, but it seems that this could eventually be compensated by a better resistance to cold.

To account for the fact that selection for weights brought about strains fitted to various ecologic hazards Leclercq assumes that

. . . genes controlling growth, size, resistance and fecundity were already submitted to natural selection and more or less linked together in the earlier history of the initial population. The latter was certainly not exceptional on account of its variability: other populations examined by previous authors exhibited a similar heterogeneity of the same characters. It seems, therefore, that most natural populations of *Tenebrio molitor* carry similar incipient strains with different ecological values. These are maintained through the life of the species not only because of continuous inbreeding but also mainly because changing ecology keeps them in co-existence in continuously changing populations.

2. *Gnathocerus cornutus*

Only two mutants have been found in this species of flour beetle which, at least in our laboratory, is difficult to rear:

(*a*) Pearl (*p*, Dyte and Blackman, 1961) is an autosomal recessive eliminating the black pigment of the ommatidia but appearing spectacled because of the black-pigmented ocular diaphragm which can be seen through the crystalline compound eye. As in *Tribolium* and *Latheticus* this mutant is identifiable in the larva and pupa by the absence of pigment in the ocelli and the ommatidia (Dyte and Blackman, 1961). Ho (1962b) reported finding an allele of *p* in descendants of a sample derived from an Oakland, California, flour mill in 1960 and 2 years later a second sample derived from the same flour mill contained a few pearl imagoes. Shaw (1965) has found that in *p* adults the Malpighian tubules are pale or transparent, while in normal beetles they are black or dark brown.

(*b*) Light ocular diaphragm (*lod*, Sokoloff and Ho, 1962) appeared spontaneously in the Berkeley pearl stock. This gene, as in *T. castaneum* and *T. confusum*, blocks the synethesis of the black pigment from the ocular diaphragm, and in combination with pearl it produces a beetle with a uniformly crystalline compound eye (Sokoloff and Ho, 1963).

These two mutants are not linked and can serve to identify groups II and III of the ten possible linkage groups in this species of flour beetle (Snow, 1962).

3. *Latheticus oryzae*

Nine mutants are known in this species of flour beetle.

a. SEX-LINKED

(1) Red (*r*, Sokoloff, 1959). A recessive of complete penetrance, good viability, and somewhat variable expressivity. Spontaneous in a cross of two pearl heterozygotes. Identifiable in *r* larvae, since the ocelli may not be visible, while in the normal these structures are pigmented black. Identifiable as soon as some pigment forms in the ommatidia in 1-day-old

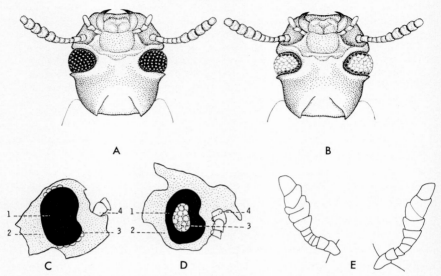

PLATE 16. Two mutants in *Latheticus oryzae*. (A,B) Normal and pearl eye, external view (prepared by Kenji Toda). Note the dark marginal ommatidia in the pearl mutation resulting from the fact that the black ocular diaphragm, an endoskeletal structure, lies under them. (C,D) Medial view of the dissected normal and pearl eyes; 1, ocular diaphragm; 2, inner surface of head exoskeleton; 3, outline of eye facet; 4, basal segment of the antenna. (E) A pair of antennae from the "fused antennal segments-1" (*fas-1*) mutant. (All drawings except E by permission of the Publisher, *American Naturalist* and of the Editor, *Canadian Journal of Genetics and Cytology*.)

pupae, and at any age in the imago, although in old imagoes the color may darken (Sokoloff and Shrode, 1960). An allele of red (*r-1*, Sokoloff, 1964) was found in a wild-type stock derived from cultures maintained at the Pest Infestation Laboratory, Slough, Bucks., England (Sokoloff, 1964h). Pearl is epistatic to red.

(2) Truncated elytra (*te*, Sokoloff, 1959). Spontaneous in a wild-type stock. The mutant resembles the *te* mutant in *T. castaneum* (Plate 17,A). It is identifiable in the pupa. The *te* gene has semilethal effects. The location of *te* is approximately six units from *r* (Sokoloff and Shrode, 1960).

b. AUTOSOMAL, LINKAGE GROUP II

(1) Pearl (*p*, Sokoloff, 1959). Recessive of good penetrance, uniform expression, and good viability. This was the first mutant found for this

species. The normally black pigment is eliminated by the p gene from all the ommatidia but not from the ocular diaphragm, giving the eye a spectacled appearance (Wolsky and Zamora, 1960, Plate 16,B,D). The normal allelomorph appears to be unstable, at least in somatic tissues: somatic mutations in the eye of $+/p$ beetles in $+/p \times p/p$ backcrosses, and resulting in imagoes with one eye pearl, the other black, or one eye black, the other partly pearl, occur at a frequency of about 1:12,500 (Sokoloff, 1959; Sokoloff and Shrode, 1960). The p gene is used as a marker identifying linkage group II.

(2) Brown body (bwb, Sokoloff, 1963). Spontaneous in a wild-type

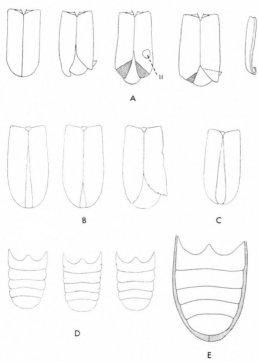

PLATE 17. Various mutants in *Latheticus oryzae*. (A) The sex-linked "truncated elytra" (te) mutations contrasted with the normal elytra. In the figure "bl" represents a blister often found in the elytra of te beetles. (B) Three examples of the autosomal recessive tucked elytra (tke). (C) The phenodeviant droopy elytra. (D) The creased abdominal sternites (cas) mutant. (E) The elongated elytra (ele) mutant. (A and D by permission of the Editor, *Canadian Journal of Genetics and Cytology.*)

stock maintained at the Pest Infestation Laboratory, Slough, Bucks., England, and kindly provided by C. E. Dyte to the writer. The normally yellowish body color approximately matches 13′.OY-O (zinc orange); 15′.Y-O (ochraceous orange) and 15′.Y-Ob (ochraceous buff), Plate XV of Ridgway's (1912) Color Standards. The *bwb* mutant appears close to 17′.O-Y (yellow ochre), 15′.Y-Oi (ochraceous tawny), 17′.O-Yi (buckthorn brown), and russet (13′O-Y-ok) on Plate XV, and Mikado brown (13′.OY-O) on Plate XXIX of the same work by Ridgway. To the unaided eye the imago appears grayish-black or sooty. The mutant can be identified in the larva by the darker (almost black) color of the urogomphi and the mouth parts. Teneral adults can be distinguished from wild type by the development of pigment in the striae which is absent in young beetles with normal body color. There is a suggestion that *bwb* produces a dark pigment in some internal tissues: pearl, non-*bwb* beetles have eyes which are crystalline over the ocular foramen, but in *bwb* beetles the same area becomes brownish-black and as dark as the exoskeleton. However, the *bwb p* phenotype may be identified from *bwb* non-*p* by casting a shadow over the eye. Viability of *bwb* is reduced 30–40% (Sokoloff, 1964g).

(3) Creased abdominal sternites (*cas,* Sokoloff, 1963), is a recessive of variable expression and incomplete penetrance. It causes the formation of a diagonally oriented groove on the abdominal sternites. The first apparent abdominal segment is more frequently affected, sometimes asymmetrically, but grooves may also appear in as many as four segments (Plate 17,D) (Sokoloff, 1964g).

Linkage data suggest these three genes are arranged in the order *bwb–cas–p.* The approximate distances between them are 36 units for *cas* and *p,* and the same distance for *bwb* and *cas* (Sokoloff, 1964g).

c. Autosomal Genes Whose Linkage Relationships Are Yet to Be Established

The following genes have recently been found:

(1) Tucked elytra (*tke,* Sokoloff, 1963). Spontaneous in brown body crosses. In a single-pair mating one male and two females appeared whose elytra were folded under as in the sex-linked *te.* This proved to be but one expression of the recessive *tke.* In other pupae the elytra may be widely separated and/or short exposing the tarsi of the hind legs. In the adult the phenotype is also quite variable, ranging from beetles with elytra short as in *T. castaneum sh,* to elytra slightly split, to elytra split starting at the scutellum, to elytra split and tips reduced

TABLE 17

Fusions in the Antennae of *fas-1* in *Latheticus oryzae*

Males		Females	
Right	Left	Right	Left
6–7	0	8–9	8–9
8–9	0	8–9	0
8–9	0	0	4–5
0	6–7	8–9	8–9
6–7, 8–9, 10–11	0	0	8–9
6–7	0	8–9, 10–11	6–7, 8–9, 10–11
4–5	0	4–5	0
4–5, 8–9	0	0	4–5, 8–9
5–6, 8–9	8–9	8–9	8–9
0	8–9	8–9	7–9
4–5	0	4–5	0
8–9	6–7, 8–9	4–5	4–5
5–6	0	8–9	8–9
4–5, 8–9	8–9	4–5	8–9
8–9	8–9	0	8–9
4–5	0	0	9–10
0	7–8	8–9	0
0	7–9	0	4–5
0	7–8	0	8–9
0	8–9	8–9	8–9
8–9	8–9	0	7–9
4–5, 6–8	4–5, 8–9	0	7–9
0	6–7	4–5	0
0	8–9	5–7, 8–11	0
4–5, 8–9	8–9	8–9	8–9
4–5	0	8–9	8–9
8–9	4–5, 8–9	4–5	8–9
0	8–9	0	6–7
0	8–9		
8–9	0		
0	5–8		
0	8–9		
0	8–9		
0	4–5		
4–5	0		
8–9	8–9		
8–9	8–9		
5–6	0		
8–9	4–5, 8–9		
8–9	8–9		
8–9	8–9		

in width; to elytra with tips tucked under in *te*-like fashion (Plate 17,B). Any combination of two of these phenotypes may occur, and the elytra may bear a blister. Viability may be reduced, but so far *tke* has not been freed of *bwb* which by itself reduces viability, so that the reduction in viability of *tke* may be due to the *bwb* gene (Sokoloff, 1965c).

(2) Droopy elytra (*dre*, Sokoloff, 1960). As in *T. castaneum* and *T. confusum* this gene causes the elytra to diverge starting often at the scutellum, and to droop at the sides of the abdomen (Plate 17,C). Probably a phenodeviant since *dre* × *dre* crosses produce predominantly beetles with normal elytra (Sokoloff, 1965c).

(3) Fused antennal segments-1 (*fas-1*, Sokoloff, 1960). Spontaneous recessive found while attempting to establish the mode of inheritance of *pearl*. The *fas-1* gene has incomplete penetrance: the stock has a very large number of beetles with both antennae normal. Even in one individual the antennal segments on one side are normal and those on the other are variously fused. Most frequently encountered are fusions in segments 8–9 (in the club). However, fusions of these segments together with fusions in segments 4–5, and 6–7 (in the funicle), or fusions of only these segments in the funicle are not rare, as can be seen in Table 17 (Sokoloff, 1965c).

(4) Elongated elytra (*ele*, Hoy and Sokoloff, 1964). A recessive of incomplete penetrance. The elytra extend beyond the tip of the last abdominal segment, but they are not elongated to the same extent as in similarly named mutants in *T. confusum* and *T. castaneum*. (compare

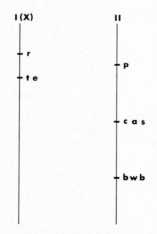

Fɪɢ. 3. Linkage maps in *Latheticus oryzae*.

Plate 17,E with Plate 14,O), the elongation being generally much less than the length of one abdominal segment. A very large number of larvae in the *ele* stock exhibit prothetely. Sometimes the elytra are expanded, becoming as long as the combined length of the two wing-bearing segments, plus the first two abdominal segments. In other cases the buds are small lateral projections. If so, only the elytral buds or the membranous wing buds or both may be everted. It is not clear whether this prothetelous condition is associated with the *ele* gene, or whether it is due to the release of quinone by the contemporary imagoes confined in the same creamer. Whatever the reason, the incidence of prothetelous larvae in this stock is far greater than any so far observed in handling other stocks or in various crosses (Sokoloff and Hoy, 1965).

d. LINKAGE MAPS IN *Latheticus oryzae*

The maps for the two linkage groups established for this species, given in Fig. 3, are based on the following values (from Sokoloff and Shrode, 1960; Sokoloff 1964g).

I	II
$r-te = 6$	$p-cas = 36$
	$bwb-cas = 36$
	$p-bwb = 50$

H. CHRYSOMELIDAE

1. *Gonioctena variabilis*

Bateson (1895) analyzed the color variation in populations of this chrysomelid living on *Spartium retama* in southern Spain. He found that color is to a great extent dependent upon sex, the males being generally red with spots, whereas the females are generally greenish with stripes on the elytra, although every color sometimes can be found in each sex. Several thousand beetles were examined in regard to spotting or striping and background color on the elytra, color of the thorax, and of the undersides of the beetles. The distribution of the spots and stripes and the ground color of the elytra in the two sexes may be given as an illustration of this sexual dimorphism: 80% of the males were of the spotted class and 18% were striped, while among the females only 25% belonged to the spotted phenotype, and 71% were of the striped class. Of the spotted beetles 73% of all males and 19.6% of all females had red as ground color of the elytra, while of the striped ones 19% of all males and 65.7% of all females had a greenish-gray as the ground color.

2. Phytodecta variabilis

Zulueta (1925, 1932) states that some color patterns in this flour beetle were determined by genes transmitted through both the X- and Y-chromosomes. The patterns included striped, yellow, red, and black, and two others not investigated. The various patterns were "found together but in different proportions in different localities. The distribution of patterns between the sexes also varies greatly. In Madrid the striped pattern is almost confined to the females, while in Granada it occurs with almost equal frequency in both sexes."

3. Melasoma (= Lina) scripta

McCracken (1905, 1907) studied two phenotypes found in natural populations of this beetle: spotted-brown elytra and black elytra, and a mutant which appeared with head, thorax, and elytra totally black. In the normal beetles the thorax is black in the center, and this area is surrounded by a wide brick-red margin. She determined by crossing these various forms that the gene S determines the spotted-brown phenotype; the gene B the black-elytra phenotype, and the AB combination the totally black phenotype. The order of dominance is $S > B > AB$. However, McCracken states (1907, p. 237) that "there is no general adherence to Mendelian proportions in the behavior in inheritance of the sport AB in Melasoma scripta but it behaves as a Mendelian recessive in first crosses with both S and B, breeding true from hybrid S."

It may be pointed out that this lack of adherence to Mendelian proportions may have been the result of faulty experimental design: she sexed the beetles by size and she mixed half broods while rearing (see details in McCracken, 1905). Hence, her work would bear repetition.

4. Gastroidea dissimilis

The same investigator (1906) investigated the inheritance of color in this chrysomelid. She found that in this beetle black is dominant and green recessive.

5. The Genus Leptinotarsa

Tower (1903, 1906) studied colors and color patterns in various "species" of chrysomelid beetles in this genus. Most of the characters chosen in various parts of the body of the beetles exhibited "continuous variation." In some cases, however, extreme variations were found: *melanothorax, multitaeniata, ribicunda,* and *angustovittata* are forms found in nature, they were given species names, and they bred true to type in the laboratory. Tower (1906, p. 92) was able to get several variations similar to these in *L. decemlineata; L. pallida, defectopunctata, minuta, melanicum,* and *rubrivittata* were found in nature or in experiments and were said to breed true to type. Intensive experiments regarding various environmental factors as agents modifying the expression of body color or pattern in *L. decemlineata* led Tower to conclude (1906, p. 209) that "temperature and moisture are the two prime factors in the production of color changes, and of these moisture is the more important." Similar experiments with *L. signaticollis* failed to elicit these modifications. Experiments with additional species led to the conclusion that "species like *multitaeniata* and *undecemlineata,* which are variable in nature, are likewise variable in experiment; while those which are stable in nature are also stable in experiment" (Tower, 1906, p. 211). Selection experiments to obtain albinic and melanic races were effective, but upon relaxation of selection a shift of the mode toward that characteristic for the species occurred (Tower, 1906, p. 266). That part of Tower's work in which he could produce mutations or their phenocopies in accordance with the way he treated his animals with abnormal conditions has never been verified (Goldschmidt, 1938).

Tower's work culminated (1910a,b) with a study of dominance of various genes in intra- and interspecies (probably today considered as races of the same species) crosses for several generations. He used *Leptinotarsa signaticollis* (LS), a species found in the foot of the escarpment on the western side of the Mexican plateau; *Leptinotarsa undecemlineata* (LU) confined to the savannahs and lower foothills from Tampico, Mexico, southward to Costa Rica and Panama; and *Leptinotarsa diversa* n. sp. (LD), similar to LS but limited to the higher foothills on the border of the Mexican plateau. The contrasting characters studied were as follows: in LU the elytra have deep greenish-black longitudinal stripes edged with a double row of punctations, the ground color being white, while in LS the punctations are present, the stripes are absent, and the ground color is grayish. The LD third instar larvae are white without stripes and the LS yellow with tergal stripes. In LD the elytra have longitudinal stripes of greenish-black edged with an irregular double row of punctations and the larvae resemble those of LS. Elytral stripes of LD and LU is a dominant character to lack of stripes in LS, yellow is dominant to white, and the tergal spots dominant to absence of spots. The material when crossed often gave almost perfect Mendelian results, and in some cases perfect ratios of 1:2:1 are obtained in the second hybrid generation, while other crosses, brothers and sisters of the same material, did not give the same, but on the contrary, quite different results. It was these anomalous results which led Tower to undertake a very large number of experiments involving several "species" of *Leptinotarsa*. Since these experiments are too detailed, only two involving LS and LD, will be mentioned here to illustrate his work. (See also, Appendix.)

(*1*) LS ♀ × LD ♂, *progeny reared at high day and night temperature and relative humidity.* The hybrid F_1 were of two phenotypes, one resembling LS and the other intermediate between the LS and LD stocks (LS-LD). The LS F_1 bred true for six generations. The LS-LD gave an F_2 in the ratios 1 LS: 1 LD:2 LS-LD. LS and LD bred true for several generations. LS-LD continued to give the same 1:2:1 segregation.

(*2*) LS ♀ × LD ♂, *progeny reared at low night temperature and high relative humidity, but high day temperature and low relative humidity.* The F_1 were invariably LS-LD, and the $F_1 × F_1$ crosses, repeated eleven times, gave 1 LS: 2 LS-LD: 1 LD. Tower suspected that differences in prevailing environmental conditions were responsible for the differences in the results of the crosses. Hence, another experiment was performed using the same parents to produce an F_1 successively at the two environmental conditions. The brood reared under the first conditions again gave 1 LS: 1 LS-LD, while the brood reared under the alternate set of conditions gave 1 LS:2 LS-LD:1 LD.

Tower concluded that "the determination of dominance and the ensuing type of behavior is clearly a function of the conditions incident upon the combining germ plasms." His later work involved rearing two or three different "species" in experimental plots with different ecological conditions which largely corroborated his laboratory observations.

I. BRUCHIDAE

Bruchus (= Callosobruchus) quadrimaculatus

Breitenbecher (1921, 1922, 1925, 1926) found a score of heritable variations in the four-spotted cowpea weevil. Some of these involved elytra and body color and they appeared to be controlled by a single, multiple allelic system. According to Breitenbecher's (1922, pp. 10–11) summary: . . . "the order of dominance of these four-color factors was red, black, white, and tan (the wild type). The females and males of the wild type have tan elytra and bodies while the females of the white mutant cultures have white elytra and bodies and the males have tan elytra but gray bodies. The black mutant stock shows a more pronounced dimorphism, since its females had black elytra and bodies, and its males tan elytra but grayish black bodies. A lesser difference is seen in the red mutation, for its females had red elytra and bodies, while its males had tan elytra and reddish-gray bodies. It is evident from this description that every male had tan elytra. In conformity with an allelomorph series, the following formulae were designated by the author: red (RR), black (R^bR^b), white (R^wR^w), and tan or wild (rr). These formulae also indicate the order of dominance." In the same paper Breitenbecher described thirty-one cases of elytral mosaic females. These females instead of possessing symmetrically red, black, white, and tan elytra, had red-black, black-red, black-tan, tan-black, and white-tan elytra. They proved to be somatic mutations.

An apterous sex-limited, autosomal recessive mutation, seen only in females and resulting in a marked lowering in viability, was described (Breitenbecher, 1925). This mutation had elytra completely missing or appearing as vestigial remnants at the wing bases.

The same investigator (Breitenbecher, 1926) gave a brief resume of his previous work and mentioned a number of other mutations he had found in these weevils. These included macula, a dominant trait characterized by the appearance of a faint bilateral spot on the elytra visible only in the male. Another gene, expressed only in the female, produced a red pronotum (the normal being black). It is stated that when insects from this pure line were mated with normal insects, the male progeny had normal black pronota. Presumably the pronota in females were red.

Some heritable variations were found having various characters affected in the two sexes, but their mode of inheritance was not worked out. These included a white eye color; longer and shorter wings than normal; modifications in the color of antennae (gray and brown), the ventrum of the body (gray or red as contrasted to the normal tan), and legs (gray versus tan); long and oval body shape; and small transparent eggs (modified from large, viscid eggs). Finally, Breitenbecher states that sex-linked lethals were also discovered.

Breitenbecher's publications cease at this point. In many of his recorded observations, data making it possible to follow what kinds of crosses he made or the numbers of progeny he scored, are not given. Nevertheless, since some of these variations have been found in *Tri-*

bolium and other beetles, it is fair to say that many of the abnormalities he put on record had genetic basis.

J. CURCULIONIDAE

Anthonomous grandis

Bartlett (1964b) reports the mutant yellow (y), a spontaneous autosomal recessive eye color mutant in the boll weevil, characterized by a lightening of the normal eye color. The mutant has good viability and fertility. Penetrance is complete, but expression varies as the beetles age.

Bartlett (1964a) describes two additional mutants in this species:

(1) Milky (m) is an autosomal recessive eye color mutant. The eyes are lighter in color than in the normal weevils.

(2) Bashful (b) is characterized by the head being recessed into the prothorax; only the snout protruded and the antennae were also deformed. The inheritance of b has not been adequately determined, but the preliminary evidence indicates that a single recessive gene controls the trait.

A recent report by Bartlett (1965) adds the following genetic information for this species.

Pearl (p): recessive autosomal gene controlling eye color. Color of eyes translucent white in adults of all ages. Expressed in pupae at the time when darkening of eyes occurs in wild-type stocks. Complete penetrance and good expressivity. Never overlaps wild type. In this stock the center of each eye is pearl-colored but the outer margin is dark giving the weevil a spectacled appearance when viewed from the side. Obtained in F_2 of an irradiation test. Probably radiation induced. Not linked to yellow, slate, or ebony. Allelic to apricot.

Apricot (pa): recessive autosomal gene controlling eye color. Allelic and dominant to pearl. Isolated in the F_2 of a cross between yellow and pearl. Eyes of the mutant show the dark ocular ring of pearl with the apricot colored ommatidia.

Ebony (e): semidominant autosomal gene controlling body color. Identical in color and gene action to slate; however, crossing results show that these are two different loci. The heterozygote has a dark bronze color. $+/e$ is distinguishable from $+/s$ even though the homozygotes are indistinguishable.

Slate (s): semidominant autosomal gene controlling body color. The homozygote is deep black color, the heterozygote is light bronze and easily classified when compared to wild type or slate. Preliminary observations on viability and fertility indicate that the heterozygote is more fertile than either homozygote and may be somewhat faster in development.

Gnarled (g): a recessive lethal mutation, presumably autosomal, but this has not been tested adequately. So far no obvious phenotype is observed in the heterozygotes carrying a gnarled gene, but the expression of the homozygote is very striking. Expression of g/g is variable but the normal appearance shows the legs and elytra twisted and shortened. The tibia is shortened and the tarsi are folded away from the body. The beak is shortened and the antenna shortened and twisted. The homozygotes sometimes eclose but are always too crippled to eat or walk and soon die. A majority of homozygous pupae darken but they do not eclose. Goodness of fit tests have run on the proportions of gnarled in the line and show a good fit to the expected 3:1 ratio for a recessive gene.

VI. Peculiarities of the Genetics of Tribolium

A. TYPES OF GENES ENCOUNTERED IN *Tribolium*

The whole gamut of sex-linked and autosomal genes encountered in other organisms may be encountered in flour beetles. Most of the sex-linked and autosomal genes so far described in *T. castaneum* are recessives; a few are incompletely recessive since in the F_1 a few individuals exhibiting slight expressions of the character may be found; a few, for example, the autosomal *b* and *sa-2* and the sex-linked *py*, are semidominant; and about 6% (if the four other dominant alleles of *Sa* are excluded) are dominants with recessive lethal effects. It is worthy of note that in *Fta*/+ × *Sa*/+ crosses a dominant synthetic lethal is produced, since *Fta Sa* beetles fail to appear among the progeny (Sokoloff, 1964a). *Fta* and *Sa*, although both dominants with recessive lethal effects and in linkage group VII, are not allelic (see linkage maps for *T. castaneum*).

Perhaps unusual is the *py* gene in *T. castaneum*. This gene is sex-linked and semidominant, drastically reducing the size and viability in both sexes and fecundity in females. It seems that in *Drosophilia melanogaster* there is no single gene discovered, so far, which reduces body size to the extent that *py* does in *T. castaneum*.

B. MUTATION RATE

The mutation rate has not been studied formally in any species of *Tribolium* or related species. Nevertheless the facts that most of the mutants have been found, for the most part, in unirradiated material; that some 40 days elapse before all the offspring of a single-pair mating complete their development; and that over one hundred mutations have been discovered in *T. castaneum* alone in a period of 6 years without a particularly intensive effort in searching for mutants, indicate that the mutation rate is high.

Certain mutants, for example, black or the sex-linked red in *T. castaneum*, have been found in various laboratory strains or in "natural" populations in feed sacks or in flour mills. These genes produced marked body or eye color changes, and examination of fairly large numbers of beetles will inevitably lead to their identification if they are present.

Sometimes more than one visible mutant has appeared spontaneously in the same individual, for example, the linked genes *ju* and *ctp* in *T. castaneum*, and the mutations *sc* and *umb* in *T. castaneum* and *T. confusum*. At times the same type of mutation has been found at about the same time in different species, for example, the genes *r* and *te* in *T. castaneum* and in *Latheticus oryzae*. A possible explanation might be that the investigator (in this particular case the writer) on finding one particular mutation in one species was alerted to the possibility of the existence of a similar mutation in another species. More difficult to explain away is the time factor involved in the discovery of the two possibly homologous genes in these two species of beetles.

There is one other case that should be placed on record. DuWayne C. Englert, working at Purdue University, Lafayette, Indiana, worked with a mutant called Gnarled (*Gn*) which proved to be allelic with Short antenna (*Sa*). Hence, *Gn* was renamed Sa-1 (Sokoloff *et al.*, 1963). P. S. Dawson, in our department, discovered Distorted (*Ds*) and this, too, proved to be allelic with *Sa*. Hence, *Ds* was renamed Sa-2. All three alleles are in linkage group VII (Sokoloff *et al.*, 1963). Englert found antennapedia (*ap*) in 1962 in the Sa-1 stock. At about the same time Dawson found antennapedia (*ap*^{D}) in his Sa-2 stock. Mutants *ap* and *ap*^{D} proved to be allelic and serve as markers of linkage group VIII. It would seem that the most plausible explanation for this case is one of coincidence, but some kind of directed mutation cannot be ruled out since the probability of these two independent events occurring in two separate laboratories is exceedingly small.

It has been noted in our laboratory that certain stocks have yielded a notably larger number of mutations than other stocks. The reason for this is not yet clear. There is no deliberate attempt to inbreed this material once the stock is established, and these stocks are not examined more frequently than others. A tentative explanation might be that a mutator gene is present in these stocks.

It has already been pointed out in the description of various mutants that the *p*, *sq*, and *r* genes appear to be unstable. The evidence for this is derived from individuals bearing one eye of one phenotype (say, pearl) and the other of another phenotype (say, wild type). These individuals have proved to be heterozygous for the particular gene involved, and thus represent cases of somatic mutation, which in individuals heterozygous for pearl occurs at a frequency of 1:10,000 in *T. castaneum* and (based on a smaller sample) of 1:12,500 in *Latheticus oryzae* (Sokoloff, 1959; Sokoloff and Shrode, 1960).

C. FACTORS AFFECTING CROSSING-OVER

1. Effect of Sex on Crossing-Over

Schuurman (1937) working with two autosomal genes (the recessive gene flesh-colored, *g*, and the dominant "V-groove," *B*) in *Tenebrio molitor* found that crossing-over was about equal in the two sexes. Lasley (1960a), investigating the autosomal recessive genes *j* and *spl* in *T. castaneum*, found the same recombination values (6–17%) when double heterozygous males or females in coupling or repulsion were backcrossed to the double recessive. Preliminary both-way backcrosses of *Be* $+/+$ *s* to $+s/+s$ (both genes in linkage group IV in *T. castaneum*) gave comparable recombination values (about 25 crossover units) for the two sexes (Sokoloff, 1963c). Thus, in contradistinction with the situation in *Drosophila* where crossing-over in the male does not occur except under unusual circumstances (Whittinghill, 1947) and in *Bombyx mori*, where crossing-over occurs only in the male (review in Tanaka, 1953), it began to appear that crossing-over might be equal for the two sexes in all autosomes of *Tribolium castaneum*. Studies in crossing-over for genes associated with linkage group VII proved to be an exception to this general observation. The data are more fully recorded in Sokoloff (1964b) and only a summary of the results needs to be presented here.

As Table 18 shows, the recombination fraction of any given pair of genes, whether in coupling or repulsion, is not significantly different within a sex, but is significantly different between sexes. The recombination values may differ only slightly (but significantly) between the sexes as shown by the values obtained in crosses A and B and C and D in the table, or they may differ considerably, as shown by the results of the remaining crosses in the table. They are increased in the male to the point that they approach 50%, giving the erroneous impression that the genes are not linked, while the values obtained from females clearly indicate that these genes are linked.

Since the *Tribolium* male is the heterogametic sex, these data violate Haldane's (1922) rule that where differences in crossing-over are found in the two sexes, it is the heterogametic sex in which crossing-over is lowered in frequency or is absent. Although it would be extremely diffi-

TABLE 18

Recombination Fractions between Various Linkage Group VII
Genes in Coupling or Repulsion

Cross	Recombination fraction	SE of recombination fraction
A. $++/Fta\ ble$ ♂ × ble ♀	0.0655	0.0101
B. $Fta+/+\ ble$ ♂ × ble ♀	0.0619	0.0182
C. $++/Fta\ ble$ ♀ × ble ♂	0.0212	0.0052
D. $Fta+/+\ ble$ ♀ × ble ♂	0.0245	0.0113
E. $++/Sa\ ble$ ♂ × ble ♀	0.5390	0.0353
F. $Sa+/+\ ble$ ♀ × $Sa\ ble/+\ ble$ ♂	0.2103	—
G. $++/Fta\ c$ ♂ × c ♀	0.4985	0.0193
H. $Fta+/+\ c$ ♀ × c ♂	0.4060	0.0182
I. $++/sa\ c$ ♂ × $sa\ c$ ♀	0.5086	0.0358
J. Sa-$2+/+\ c$ ♀ × c ♂	0.3953	0.0250
K. $sa+/+\ c$ ♀ × $sa\ c$ ♂	0.3915	0.0200

cult to prove cytologically, it is highly probable that this discrepancy in crossover values in the two sexes of *T. castaneum* results from a difference in the distribution of a single chiasma. Since linkage information is so far limited, it is not known whether this phenomenon, presumably associated with only one arm of the seventh linkage group in *T. castaneum*, will be found for the other autosomes. However, it is clear that in linkage studies, at least in *T. castaneum* but possibly in other species of beetles as well, it is necessary to set up crosses both ways, otherwise linkage relationships between the genes being tested might be missed.

2. Effect of Other Factors on Crossing-Over

Since other factors are known to affect gene recombination in *Drosophila*, a preliminary study attempted to determine whether such factors had an effect on the frequency of crossing-over in *Tribolium*. The factors studied were age, temperature, and irradiation on recombination between *Be* and *s*, found associated with linkage group IV and 25 units apart. The data from Sokoloff (1963c), summarized in Table 19, are somewhat unsatisfactory, but it seems that, at least for these genes, there is a slight (but significant) increase in crossing-over in the male (to about 27%) resulting from age and X-ray effects; on the other hand, cold had no effect in the male, but reduced crossing-over significantly in the female (to about 22%). These experiments would bear repetition (since numbers

TABLE 19

Crossovers and Totals Observed at Four-Week Periods under
Various Experimental Conditions in *Tribolium castaneum**, †, ‡

4-Week period	I		II		III		IV	
	A	B	A	B	A**	B	A	B
1	397/1744 (22.76)	392/1612 (24.32)	263/1006 (26.14)	245/1066 (22.98)	—	—	166/616 (26.95)	43/185 (23.24)
2	450/1711 (26.30)	162/634 (25.55)	130/515 (25.24)	190/829 (22.92)	—	143/705 (20.28)	251/834 (30.10)	44/171 (25.73)
3	375/1511 (24.82)	72/256 (28.12)	157/656 (23.96)	33/148 (22.30)	—	172/750 (22.93)	167/668 (25.00)	12/47 (25.53)
4	238/897 (26.53)	7/37 (16.22)	138/542 (25.46)	18/114 (15.79)	—	75/301 (24.92)	8/23 (34.79)	—
5	32/94 (34.04)	—	—	—	—	—	—	—
Total	1492/5957 (25.05)	633/2539 (24.93)	688/2715 (25.34)	486/2157 (22.53)	—	390/1756 (22.51)	592/2141 (27.65)	99/403 (24.57)

* I, Controls reared at 32°C, 70% relative humidity; II, reared at 24°C; III, P_1 cold-shocked for 40 hours, F_1 reared at 24°C; IV, exposed to 3210 r X-rays.

† A = $Bes/++ \, ♂ × +s/+s \, ♀$; B = $Bes/++ \, ♀ × +s/+s \, ♂$.

‡ Numbers in parentheses are per cent crossovers.

** None of the beetles survived the exposure to cold.

of observed adults is relatively small) and expansion with other auto-somal genes.

D. MULTIPLE ALLELES IN FLOUR BEETLES

In this section a number of mutants in *Tribolium* found by various investigators in different laboratories have been excluded. The mutants resemble each other in phenotype (for example, *j* and *j^E*, *lod* and *lod^D*, *Mo* and *Mo-1* in *T. castaneum* and *e* and *e^{L and H}* in *T. confusum*), and they proved to be allelic when tests of allelism are performed. The residual genetic background is different and it could be argued that they are alleles rather than recurrences of the same gene. Since information on the different records of a particular mutation has been included in previous sections describing the mutants, their inclusion here would unnecessarily expand this section, which for the most part places emphasis on mutants which proved to be allelic but nevertheless were phenotypically somewhat different from other alleles at that locus.

1. *Tribolium castaneum*

a. The *Sa* Locus

This locus has now eight known alleles. Five of these (*Sa*, *Sa-1*, *Sa-2*, *Sa-3*, and *Sa-4*) are dominant with recessive lethal effects (since *Sa/Sa* die) and serve as convenient markers for linkage group VII (Sokoloff *et al.*, 1963). In addition to these dominant alleles three more are known (*sa*, *sa-1*, and *sa-2*). The first two are incompletely recessive, and the last is a semidominant. Phenotypically the dominant alleles and the homozygous semidominant and incompletely recessive genes (*sa/sa*) resemble each other in the expression of the antennae, the antennameres being variously fused. Sometimes the fusions are so pronounced that they produce a curved antenna. They differ in the expression of the legs: the *Sa* alleles seldom have the femur affected, but the tibia may appear curved, gnarled, or knotted, and the first tarsomere may be fused to the tibia; the *sa* alleles, on the other hand, suffer from considerable reductions of the femur of all legs, and if the tibiae are affected, they become curved but seldom gnarled or knotted.

It has been suggested (Dawson and Sokoloff, 1964) that *Sa* might be a complex locus. Preliminary crosses between *Sa-3* and *sa* and backcrosses of *Sa-3/sa* F$_1$ with $+$ *sa*/$+$ *sa* (Sokoloff, 1964, unpublished) have produced a few normal individuals (where none should appear if these are true alleles) indicating that a complex locus may indeed be present in the seventh linkage group. The data need to be expanded using other marker genes on both sides of *sa* to be sure.

b. THE BLACK LOCUS

If one considers all the black mutations described above as independent recurrences of the same gene and not as alleles (there have been possibly six independent occurrences of this semidominant gene which produces a black body color in the homozygote and a bronze body color in the heterozygote), there are still two mutants which are allelic to *b* but they behave differently: cordovan and tawny (b^{cd} and b^t, respectively). These, in homozygous condition, are said to resemble $b/+$. Individuals which are b/b^{cd} are darker than b^{cd}/b^{cd}, but they are lighter than b/b (Eddleman, 1964). Apparently the same is true for the F_1 of b/b and b^t/b^t crosses (Dyte *et al.*, 1965). Comparisons between b^t/b^t and b^{cd}/b^{cd} have not been possible since Eddleman has not made his mutant available (Dyte, 1964 personal communication).

c. THE RED LOCUS

Several mutations have been cited which are allelic with the sex-linked recessive eye color mutant red (r): r^H, r^D, and *r-1*. The writer has not been able to compare the first three with *r-1* Eddleman and Bell since this mutant has not been released. However, outcrossed from the stock and recovered in the F_1 or F_2, *r* and r^H do not differ, ranging from pink to red to Burgundy red color. Mutant r^D is allelic to *r* but sex-influenced, being much darker in the female.

d. THE PEARL LOCUS

Only two alleles are known at this locus on the second linkage group: pearl (p) and pink (p^{Pk}). The latter is dominant to p, producing a chestnutlike phenotype in intensely colored p^{Pk}/p or p^{Pk}/p^{Pk} individuals.

e. THE JUVENILE UROGOMPHI LOCUS

Three instances of mutants bearing urogomphilike or bristlelike appendages next to the anal opening have been discovered in unrelated stocks. The *ju* and *eju* mutants are recessive, while *rju* is incompletely recessive. Preliminary tests indicated that they were not allelic. Recent tests, however, indicate that *rju* and *eju* are found in the same linkage group as *ju*. Furthermore, *rju* and *eju* give very nearly the same recombination values with *fas-2*. The location of *rju* and *eju* in respect to *ju* is being investigated by means of three-point crosses.

f. The Antennapedia Locus

Three alleles are now known: *ap*, *ap^D*, and *ap^s* (described above as fused antennal segments-6, *fas*-6).

g. Other Loci

Mutants *fas*-3, *aa*, and *dfl*, affecting the antennae, elytra and/or legs, have been cited above.

2. *Tribolium confusum*

Although pearl has been discovered four times, black at least three times, and ebony twice in different laboratories or strains, they are here regarded as recurrences of the same mutation (except for *b*-3, see above). The only allelic series known in this species is that involving eyespot.

a. The Eyespot Locus

McDonald's *es* mutation, while easy to identify in the pupa or teneral adult, is difficult to identify in the aged imago, requiring intense illumination and shading to detect a small, dark reddish area in an otherwise black eye. A recessive allele of *es*, *es^lt* produces a uniformly red eye (see description of these alleles, p. 81).

b. The Thumbed Locus

The two alleles, *thu* and *thu^s*, have already been cited on p. 101.

3. *Latheticus oryzae*

The Red Locus

Two independent findings of *r* have been made by the writer in somewhat related stocks: The original wild-type stock existed at the Pest Infestation Laboratory, Slough, Bucks., England for many years but its origin is obscure. About 1953 Dr. L. M. Roth requested and received a sample of *Latheticus*, and this wild type produced a pearl mutation which he furnished to Dr. T. Park, who in turn gave these strains to the writer in 1955. The wild-type strain was contaminated with pearl and by appropriate crosses a wild-type strain free of this mutant was established. The first *r* mutation was found in some single-pair matings between the + and *p* strains. A search for *r* in the established + strain gave negative results so that *r^+* must have mutated at the time it was found. A sample of wild type obtained from the same laboratory in England in 1963, when crossed with *bwb*, produced red beetles in several single-pair matings indicating the *r* gene had been maintained in the wild-type strain for some time. Both reds appear to be the same in phenotype. Thus, it is possible that the two reds, *r* and *r-1* are independent occurrences of the same gene or they are allelic.

E. AN UNUSUAL MODIFIER-SUPPRESSOR SYSTEM IN *Tribolium*

A sex-linked gene modifying the phenotype of the similarly sex-linked *r* gene has been found in *T. castaneum*. The supporting data are in press (Sokoloff, 1965a) so that only a summary is reported here. In crosses involving two reds, exceptional females bearing normal (black) eyes were produced. These females, mated to red males from the stock again produced exceptional black-eyed females and red-eyed females in equal numbers. Their sons were all red-eyed, indicating that the mothers were indeed *r/r* but that the red phenotype was being suppressed in half of the females. A closer examination of the male progeny of these females revealed that they were of two eye phenotypes: light and dark red-eyed in about equal numbers. A very large number of crosses with all sorts of mating combinations was set up, and the results are summarized in qualitative fashion in Table 20.

The upper block shows the pattern of inheritance of red without the modifier, and the middle block the results of identical crosses when the modifier is included. The only difference is that in crosses *i* and *j* a few light red males are produced as a result of crossing-over between *r* and M^r and their wild-type alleles, and since *r* has a variable expression (especially as a function of age) the few light eye color crossovers might be missed. The effects of the M^r gene cannot be missed when beetles represented in the first two blocks are crossed in the manner shown in the last block in the table. When two reds are crossed, the expectation is all red-eyed progeny. Instead, black-eyed females are obtained in crosses *k* to *n*. Crosses *m* and *n* are also peculiar in that the type of male used *apparently* determines the phenotype of half of his daughters. Thus, in cross *m* the light red-eyed males engender the production of light red-eyed daughters, whereas in cross *n*, the mating of the same females with dark red males produce dark red-eyed female offspring. Matings of light red males with light red females or dark red males with dark red females result in progeny which are either light red or dark red, indicating that the first females are *r/r* and the second females are rM^r/rM^r. Once it is understood that *r* and M^r are sex-linked, there is no difficulty in the interpretation of the genetic results.

TABLE 20

Phenotypes Produced in Various Crosses Involving Red (r) in the Absence and in the Presence of the Modifier (M^r)

Cross	Black-eyed		Light red-eyed		Dark red-eyed	
	♂	♀	♂	♀	♂	♀
a. $r/r \times r/$			+	+		
b. $+/+ \times r/$	+	+				
c. $r/r \times +$		+	+			
d. $+/r \times +$	+	+	+			
e. $+/r \times r$	+	+	+	+		
f. $r M^r/r M^r \times r M^r/$					+	+
g. $++/++ \times r M^r/$	+	+				
h. $r M^r/r M^r \times ++/$		+			+	
i. $++/r M^r \times ++/$	+	+	+		+	
j. $++/r M^r \times r M^r/$	+	+	+		+	+
k. $r M^r/r M^r \times r +/$		+			+	
l. $r +/r+ \times r M^r/$		+	+			
m. $r M^r/r + \times r +/$		+	+	+	+	
n. $r M^r/r + \times r M^r/$		+	+		+	+

What is difficult to interpret is the physiological process involved. The red pigment produced by the r gene in *Tribolium castaneum* is apparently an ommochrome, since chromatographic and UV analyses fail to demonstrate the presence of pteridines (Sokoloff, 1961, unpublished). The modifier and the wild-type allelomorph must also control ommochromes.

If only the male data were available, it would be simple to represent the order of these genes as $r \to rM^r \to +$, the modifier producing an intermediate color between red and black.

The data from females fall into line to the following extent:

1. r/r produces a red as light as r.
2. $+/r$ produces a black eye as expected if r is a recessive.
3. rM^r/rM^r produces as variable a dark red eye as rM^r but (and here the interpretation of the results becomes difficult):
4. $rM^r/r+$ instead of producing an eye color intermediate between r/r and rM^r/rM^r produces a black pigment in the eye which appears identical to that produced by $+/+$, $+/r$, $+/M^r$, or $++/rM^r$.

Insofar as is known to the writer, this situation is unique. A review of the literature on eye color modifiers in other insects fails to show a similar situation in the Diptera (primarily *Drosophila*), Lepidoptera, and Hymenoptera. In all these organisms modifiers either lighten or darken the eye color phenotype, but the darkening process apparently never attains the black color characteristics of wild type. The modifier of red in *T. castaneum*, on the other hand, while behaving as a modifier in hemizygous or homozygous condition, darkening the red phenotype, is capable of producing a black-eyed female when its genotype is $rM^r/r+$. The only explanation that can be offered for this phenomenon, lacking any biochemical analysis of the nature of the black pigment in $rM^r/r+$, is that the interaction of r and M^r in heterozygotes somehow results in the formation of black pigment.

VII. Time of Gene Action

Coleoptera differ from Diptera in that the larva possesses essentially the same mouthparts as the adult, and they differ from coarctate Diptera or obtect Lepidoptera in that the pupal stage is exarate, i.e., bare. These differences have made it possible to determine the time of action of certain genes. For example, in the wild-type larva the ocellus is a black-pigmented structure. Genes modifying the pigment of the adult eye modify the pigment of the ocellus. Thus, the larva emerges from the egg with pigmentless ocelli in the pearl mutant.

The various mutants producing a red eye in the adult, as a result of the action of the sex-linked r, es, es^{lt} genes, or the autosomal recessives. The c, p^{pk}, rus, etc. genes in *T. castaneum* and *T. confusum*, are distinguishable from the wild type in the larva either by having no detectable ocelli, or ocelli clearly distinguishable from the normal depending on the color and intensity of the pigment in the larval eyes.

Genes which affect the morphology of the eye, such as Bar eye, microcephalic, Microphthalmic, squint, and glass, produce no changes in the larva which could be detected with the dissecting microscope, but they become evident as soon as the compound eye has developed sufficiently to become visible through the pupal skin. The same is true for the deposition of pigment in the ocular diaphragm: in *lod⁺ p* beetles black pigment is deposited in this structure late in the pupa in *T. castaneum*, *T. confusum*, and *Gnathocerus cornutus*, and the imago generally emerges with a fully pigmented ocular diaphragm. In *Latheticus oryzae*, on the other hand, deposition of pigment in this structure begins in the pupa but is completed after the imago emerges from the pupa.

Many genes affecting the walking legs, antennae, or other appendages in the adult already have visible effects as soon as the larva emerges from the egg. Perhaps the most striking are the effects produced by the homeotic mutants antennapedia (*ap*) in *T. castaneum* and labiopedia (*lp*) in *T. confusum*. In *ap* the antenna is modified into a leglike structure in the adult and in the larva. The legs and antennae differ in morphology in these two stages so the modified antenna in *ap* larvae resembles the larval legs and in the adult the adult legs (Plate 7, H–K). The labial palps differ somewhat in the normal larva and adult, but not to

145

PLATE 18. A: Phase-contrast photograph of the posterior end of the normal larva showing the urogomphi.

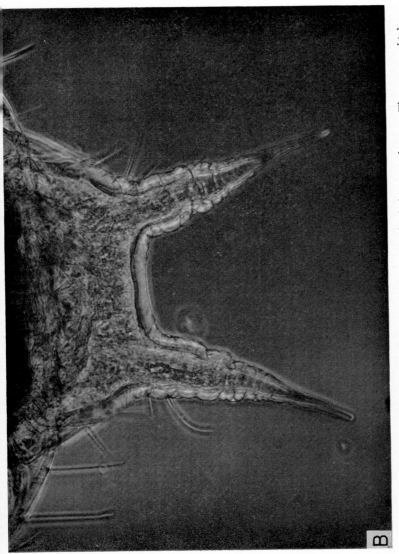

PLATE 18. B: Phase-contrast photograph of the posterior end of the normal pupa. The urogomphi have the same appearance in the two sexes.

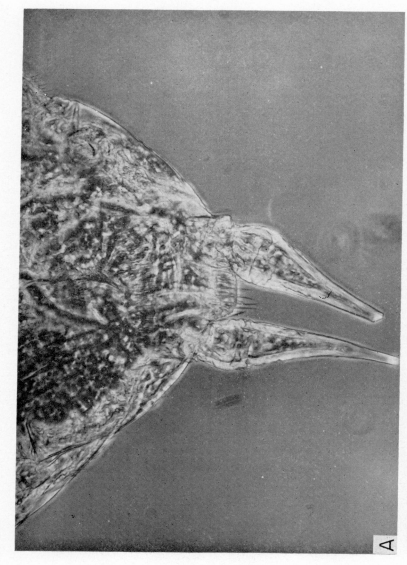

PLATE 19. A: Phase-contrast photograph of the posterior end of the *ju* male showing that the urogomphi retained in the adult can be sizable structures.

PLATE 19. B: Phase-contrast photograph of posterior end of the *ju* female showing the same appendages. In both sexes the urogomphi originate near the anal opening.

the extent that the legs differ in the two stages. In *lp* larvae the labial palps are replaced by typical larval legs and in adults by typical appendages resembling most the first pair of walking legs (Plate 11, E–I). The transition from a larval to an adult antennal leg in *ap*, and from a larval to an adult labial leg in *lp*, is evident as soon as the pupa forms, so that these pupal appendages must form within the cuticle of the larval appendages, and they extend at the pupal ecdysis.

There are some genes in *Tribolium* which seem to affect only pre-imaginal stages: extra urogomphi (*eu*) produces one or two extra urogomphi in the larva and the pupa. What changes occur in the adult is open to further investigation.

Of particular interest are the mutants designated as *ju, rju,* and *eju* in the fourth linkage group of *T. castaneum*. These genes cause the retention of urogomphilike structures in the adult. Sometimes, as in *ju*, these structures are of considerable size (Plates 5, R,T,U,V) and other times, as in *rju*, these structures are bristlelike. Regardless of their size, they form next to the anal opening in both sexes. The writer (Sokoloff, 1963b) at first thought that they were atavistic structures since (when they are strongly expressed) they appear like cerci of lower orders of insects. Examination of these structures in the adult with the aid of a phase-contrast microscope and comparisons with pupal urogomphi reveal that these appendages in the two stages are very nearly identical (Plates 18 and 19). A more reasonable explanation is that these cases constitute examples of paedomorphosis (retention of juvenile stages in the adult) not previously recorded in the Coleoptera.

One more mutation should be included here: *sta* (spikes on trochanters and antennae). The *sta* mutation was first detected because of a chance retention of huge processes arising from the trochanters of all legs and from the second segment of the antenna in the adult. The stock that has been established reveals that the *sta* gene has its effect primarily on the legs and/or antennae of the larva, since about 80% of the larvae bear some kind of deformity on these appendages. At metamorphosis the fleshy processes are resorbed and only a few individuals in the imago exhibit the spikes on the trochanters and the antennae (Plates 10, A–D).

Finally, it may be noted that the fact that *Tribolium* lives in a particulate medium from which the eggs, larvae, pupae, and adults can be extracted with ease makes it possible to determine the time of action of various lethals. This has been done particularly for the dominant mutations having recessive lethal effects and it has been inferred that individuals bearing these genes in homozygous condition, for the most part, die in the egg state (see, for example, Sokoloff and Dawson, 1963b).

VIII. Homoeotic Mutants
in Tribolium

Several cases of hereditary homoeosis have appeared in *Tribolium:* in *T. confusum* labiopedia and knobby prothorax; in *T. castaneum* antennapedia, alate prothorax, megalothorax, and spikes on trochanters and antennae. Details of these can be found in the section describing the various mutations found in the corresponding species. A few remarks seem appropriate here in regard to homoeotic mutants in general, and a comparison of homoeotic mutants in *Drosophila* and in *Tribolium.*

Examples of homoeotic substitutions in the Arthropoda are numerous. They have been well documented by Bateson (1894) and in Villee's (1942, 1945) review articles. Instances of such cases of homoeosis have been recorded for Crustacea, and among the Insecta in the Orthoptera, Homoptera, Neuroptera, Lepidoptera, Diptera, Strepsiptera, and Hymenoptera. In the Diptera the best investigated cases of homoeosis have been those resulting from genetic changes in various species of *Drosophila.* These include bithorax and bithoraxoid, aristapedia, proboscipedia, tetraltera, ophthalmopedia, podoptera, and extra sex combs (see Villee, 1942, 1945, for other references).

Various beetles found in nature and reported as teratologies also exhibit homoeotic modifications. These have been reviewed and the literature cited in Daly and Sokoloff (1965). The two best investigated homoeotic mutations in *Tribolium* are antennapedia (*ap* and *ap^D*) and labiopedia (*lp*). The first is an autosomal recessive like all of the homoeotic mutants described in *Drosophila,* but *lp* is sex-linked, and thus, it is the first homoeotic mutant found among the Insecta which is associated with the X-chromosome.

Morphological and anatomical details associated with *lp* may be found illustrated in Daly and Sokoloff (1965) and need not be repeated except in summary form here. The labial legs in the larva and the adult *lp,* usually include the apex of the trochanter and all the parts of a normal leg distal to the trochanter. Muscles corresponding to those which insert distal to the coxa, but very slender and few in number, may be found in the labial legs. However, these do not move *in situ* and are often

151

broken off by autotomy. Other structures (nerves, tracheae, chordotonal organs, and glands) characteristic of the walking legs are found in the labial legs.

The *lp* mutant is the second mutant known among the insects to affect the mouthparts, the other being proboscipedia (*pb*), described by Bridges and Dobzhansky (1933) in *Drosophila melanogaster*. Both are probably mutations of genes which are involved in the early development of mouthparts. They probably have the effect of removing part of the labium-specific organ field so that the prospective labial tissue is influenced by the adjacent organ fields for antennae or legs. Mutants *lp* and *pb* are alike in the complete replacement of a normal part of the gnathal appendages by appendicular structures which do not move; in the occasional blackening of malformed parts which commonly results in insects from the action of tyrosinase and hemolymph following injury and exposure to air; in being handicapped during feeding by the labial appendages; but the beetles are much less affected in that they mature as larvae and live for weeks as adults in contrast to the few days survival of adult *Drosophila*. However, both mutations show lethal to semilethal effects.

Aside from differences in the mode of inheritance of *pb* and *lp*, the two mutants differ in the following characteristics: in *pb* only the apical portions of leglike or aristalike appendages develop while *lp* has almost fully developed limbs which are always leglike. Owing possibly to the specialized character of the feeding apparatus, other parts are drastically altered, but in *lp* only the labium is affected. The mouth structures developed in *pb* are said to resemble the biting mouthparts of more generalized insects, whereas the labium of *lp* exhibits no atavistic characteristics.

The modifications of the beetle's labium are present in all stages of postembryonic development (conforming with general observations on other mutants having the legs affected or whose eye pigment is modified by eye color genes) but those of the fly are said to be found only in the adult instar.

In contrast to the features listed by Goldschmidt (1940) or by Villee (1942, pp. 497–498) as common to homoeotic mutants in *Drosophila* or other forms, labiopedia seems to exhibit just the opposite: (1) the phenotypes are *not* variable and do not range from normal to fully formed heteromorphic organs; (2) there is *no* large variation in penetrance; (3) there is a high degree of symmetry, and (4) there is a high degree of right-left correlation in expression.

The labiopedia mutant provides striking verification of the classic concepts of insect morphology and lends support to Goldschmidt's (1940) observation that the homoeotic organ is usually a palpus which becomes variously modified. The labial palpi have long been thought to be serially homologous to legs and in *lp* the labial legs *are legs*. Recognizing that serial homology entails similarity in development, one may imagine that the prospective palp material, in the absence of palp-forming factors as a result of the *lp* mutation, comes under the influence of the adjacent organ field for prothoracic leg. This becomes evident after introduction of mutations affecting legs: individuals homozygous for *stl* (stilted legs) and hemi- or homozygous for *lp* have their walking legs *and* labial legs modified in the same way. Similar results were obtained by Waddington (1940) when he combined aristapedia with certain leg-influencing genes in *Drosophila*.

IX. Gene Homology in Tribolium Species and Related Forms

All of the flour beetles of the genus *Tribolium, Gnathocerus cornutus,* and *Latheticus oryzae* are in the subfamily UIominae, and *Tenebrio molitor* belongs to the subfamily Tenebrioninae. Hence, the various species represent various degrees of taxonomic relationship within the family Tenebrionidae. Since they evolved from a common ancestor, it is expected that they have many genes in common.

Earlier, Sokoloff and Shrode (1960) proposed that the gene pearl in *Latheticus oryzae, Tribolium castaneum,* and *T. confusum,* and flesh-colored in *Tenebrio molitor* be considered homologous. The proposal of homology for these genes was based on the following points: (1) In all species the pearl gene eliminates the pigment from the larval ocelli and from the ommatidia of the compound eye. (2) Pearl (or flesh-colored) acts in epistatic fashion to other eye color genes (except those which are allelic to pearl). (3) Pearl (or flesh-colored) appears to be unstable; it is usually the first eye mutant found in these various tenebrionids. Furthermore, somatic mutations have been recorded in *Latheticus oryzae, Tribolium castaneum, T. confusum* (Sokoloff and Shrode, 1960), and *Tenebrio molitor* (Schuurman, 1937). Dyte and Blackman (1961) upon finding pearl in *Gnathocerus cornutus* independently made the same suggestion.

Owing to the fact that interspecies hybrids cannot be obtained even among beetles belonging to the same species group, the only way homology can be established between the various genes is by finding out their linkage relationships. Since knowledge of the genetics of *Tribolium* or related forms is still rudimentary, it might be expected that little else could be said about gene homology in these forms. Admittedly, some of what follows is based on largely fragmentary evidence, but considering the relatively short period that flour beetles have been investigated it seems remarkable that such a section could be included at all in this review. At present, the writer believes that the mutants listed in Table 21 must be considered as resulting from the action of homologous or possibly homologous genes.

TABLE 21

Homologous and Possibly Homologous Mutations in Tenebrionids

Name of mutant	Gnathocerus cornutus	Latheticus oryzae	Tribolium anaphe	Tribolium castaneum	Tribolium confusum	Tribolium destructor	Tribolium madens	Tenebrio molitor
Sex-linked								
Red, red, eyespot?	—	X	—	X	X	—	—	—
Miniature appendaged	—	—	—	X	X	—	—	—
Truncated elytra	—	X	—	X	—	—	—	—
Autosomal								
Pearl (flesh-colored)	X	X	—	X	X	—	—	X
Light ocular diaphragm	X	—	—	X	X	—	—	—
Jet (ebony?)	—	—	—	X	X	—	—	—
Umbilicus	—	—	—	X	X	—	—	—
Scar	—	—	—	X	X	—	—	—
Alate prothorax, knobby prothorax?	—	—	—	X	X	—	—	—
Blistered elytra	—	—	—	X	X	—	—	—
Rough	—	—	—	X	X	—	—	—
Akimbo, thumbed	—	X	—	X	X	—	—	—
Tucked elytra, bent elytral tips	—	X	—	X	—	—	—	—
Elongated elytra	—	X	—	X	X	—	—	—
Warped elytra	—	X	—	X	X	—	—	—
Droopy elytra	—	X	—	X	X	—	—	X
Bent tibia	—	—	—	X	X	X	X	X
Sternites incomplete	—	—	X	X	X	X	—	—
Creased abdominal sternites	—	X	—	X	X	—	—	—
Reduced juvenile urogomphi	—	—	—	X	X	—	—	—
Melanotic stink glands	—	—	—	X	X	—	—	—

The easiest way to deal with this subject is by considering the sex-linked genes first. A good case may be argued for homology between the miniature appendaged (*ma*) genes found in *T. castaneum* (henceforth to be referred to as CS) and in *T. confusum* (CF). In both species this gene causes the whole body to appear shorter and more compact, but the effect is most easily observed in the legs and elytra, the latter becoming about two-thirds as long as in the normal beetles, and the podomeres of all legs becoming short and thick. Since the phenotype (if proper allowance for differences in the normal beetles is made) is nearly the same, and the two genes are sex-linked, it is quite probable that they are homologous, although it must be admitted that in CS the autosomal mutant troll appears similar to *ma* in some respects. In CS the gene *sp* (spotted) is an incomplete recessive in which certain areas of the elytra fail to become pigmented in the teneral adult but pigment develops as the beetles age. It is located about 46 units from *r* (red), an eye color mutation. *St* (Striped), in CF is a dominant with recessive lethal effects characterized by a light stripe in the middle of each elytron observable in the teneral adult. The stripe acquires pigment as the beetles age. *St* is located about 38 units from *es* (eyespot), an eye color gene. It is not clear whether *St* and *sp* are homologous. The distance between *St* and *es* in CF and *sp* and *r* in CS is of about the right order of magnitude, but if *St* is homologous with *sp* it must have an allele which does not have recessive lethal effects. Homology between *es* and *r* may be argued on stronger grounds: *r* resembles the phenotype of es^{lt}; *r* has several alleles one of which considerably darkens the eye producing a phenotype similar to *es*. However, more markers are needed to locate *es* and *St* more precisely before homology of these genes can be established.

For sex-linked genes a stronger argument may be advanced for homology between *r* and truncated elytra (*te*) in CS and *L. oryzae* (LO). The reds have a similar appearance, and the *te* mutants have the same range in phenotype. The *r–te* distance is 12 units in CS and about 6 units in LO. Despite the similarity in the distance between *r* and *te*, the two *te* genes show differences in physiological effects: *te* in CS is semilethal to lethal in effect but in LO it is only a semilethal. Considering that we are here dealing with species in different genera, the relative distances between the two genes show fair agreement. The differences in viability might be due to the fact that we have here examples of different alleles. Evidence that this might be so is the finding by Dawson

(1963a) of an allele of *te*, *te^D*, which did not have as marked a reduction in viability as the original *te* mutant in CS.

It has been pointed out (Chapter III) that CF has one pair of chromosomes fewer than CS as a result of translocation of one pair of autosomes to the X and y pair forming a neo-X and a neo-Y which in CF are considerably larger than the X and y in CS. The autosomal portion of the neo-Y has undergone heterochromatinization and thus, whatever genes were carried in that autosome have become inert, the neo-Y serving purely a mechanical function (Smith, 1952a,b). Thus, it is expected that some genes, behaving genetically as autosomal in CS, will behave as sex-linked in CF. Of the few genes behaving as sex-linked in CF, only two resemble genes which in CS behave in autosomal fashion. In CS *ptl* (prothoraxless) in heterozygous form causes marked changes in the prothorax and the legs (see Plate 8, D,E). A sex-linked mutant has been discovered in CF which largely resembles *ptl* in CS. It has been designated as *ptll* (prothorax-less-like). (Plate 11,J). The deformities in the prothorax and the legs are quite variable and the phenotypes of *ptll* and *ptl* overlap to a large extent. However, *ptll* differs from *ptl* in that, in the latter, the effect is confined to the prothorax and the prothoracic appendages, whereas in *ptll* the gene also affects the labium, which is somewhat reduced in size and is drawn ventrally, away from the mouth. If *ptl* and *ptll* are homologous, the effect of *ptll* on the labial structures might be attributed to the close association of this gene with another which affects the labium and which, in hemizygous condition, acts as a single gene. Clearly, this point will require further investigation. In CF, *pas* (pointed abdominal sternites) is a sex-linked recessive causing a serial homology of the abdominal segments. All the abdominal segments behind the apparent first abdominal segment are formed in the shape of this segment (Plate 11,D). In CS the *ppas* (partial pointed abdominal segments) gene does the same thing, except that only the apparent second abdominal segment is formed in the shape of the first (Plate 7,O). However, recently a *ppas* beetle has been found with apparent segments 2 and 3 with median anterior processes identical to that in the first apparent abdominal segment. Thus, *pas* and *ppas* are becoming closer in phenotype. If experiments in progress prove that *ppas* and *ptl* are linked, then homologies between these genes in CS and *pas* and *ptll* in CF may be established, and evidence would be available for the pair of autosomes which, in the ancestry of CF, became associated with the X and Y of this species.

If there are difficulties in establishing homologies between sex-linked genes, these difficulties are greatly magnified when we turn to autosomal genes since linkage relationships among the various mutants are largely to be established and genes in different linkage groups may exhibit similar phenotypes. Nevertheless a modest beginning has been made and the results are quite promising. Table 21 includes a summary of the mutants for which some evidence of homology is available and others for which such evidence is lacking, but the phenotypic expression of the gene, and its mode of inheritance, suggest possible homology. A few remarks about some of these genes will attempt to convince the reader why the writer considers these genes homologous.

The gene pearl (p) has been found in two species of *Tribolium* (CS and CF), *Gnathocerus cornutus* (GC), *Latheticus oryzae* (LO), and *Tenebrio molitor* (TE). In all these species the p gene eliminates the black pigment of the ocellus or the compound eye; it is epistatic to eye color genes in CS, CF, LO, and TE, and it appears to be unstable in all these species since somatic mutations are present in heterozygotes at high frequencies in LO and CS and they have been observed in CF and TE. That they are homologous in LO and CF is evidenced by the fact that the recombination values between p and *cas* (creased abdominal sternites) for these two species are 36.32 ± 2.65 (Sokoloff, 1964g) and 37.61 ± 4.82 units (Sokoloff, 1964f), respectively, a remarkable agreement since these species are in distinctly different genera.

The *lod* (light ocular diaphragm) gene has been found in GC, CS, and CF. This gene eliminates the melanin pigment from the ocular diaphragm, an endoskeletal structure found under the peripheral ommatidia of the compound eye. No other linkage data is available for GC since only two mutants are known in this species, but p and *lod* appear to be in separate linkage groups. In CS *lod* is linked with the semidominant b (black). In CF b and *lod* fail to show linkage, but conceivably they could still prove to be in the same linkage group, the order of the genes having been altered in the course of evolution of the two species.

Homology has been proposed for jet in CS and ebony (e) in CF purely on the basis of the appearance of the body color of these two mutants. Linkage of e with other genes (except ble) in CF is yet to be established.

The mutants umbilicus and scar in CS and CF are probably homologous on the basis of phenotype. The two mutations have been found on the same individuals in the two species and they may be linked to each other, but this may be difficult to prove because of their incomplete penetrance.

Alate prothorax and knobby prothorax in CS and CF, respectively, are homoeotic mutants. Both produce processes or winglike structures on this part of the body. They may be homologous, but demonstration of homology from linkage data may prove difficult because of the low penetrance of these genes on outcrossing.

Two mutants in CS and two in CF produce blisters, and they bear the same names: rough (*ro*) is incompletely penetrant in both species producing a blister on the elytra in some pupae which remains in the elytra of the adult. The linkage relationships of these two genes are yet to be determined, but in both species *ro* individuals bearing no blisters have the elytra with roughened appearance. Two genes are designated "blistered elytra" (*ble*) in these two species. In both species the blisters seldom occur in the pupa but they appear in the adult. The *ble* gene is incompletely penetrant in both species, but to the writer the blistered condition in CF is more pronounced than in CS. There is also a difference in the size and position of the elytral blister in the two species. Linkage of *ble* has been established in CS (in the seventh linkage group) and in CF (in the fifth linkage group). Unfortunately, the other genes in these linkage groups are different, so it is not possible to state whether the same autosomes are represented in the two species.

The akimbo and thumbed mutants are given homologous status on the basis of the phenotypic expression of the elytra, which appear split or raised off the body in the adult. This condition arises from the formation of a blister in the membranous wings which displaces the elytra to a varying degree in the pupa. In the adult the blister interferes with the proper placement of the elytra over the abdomen before sclerotization, and after this process is completed the elytra remain misplaced and misshapen.

Homology is proposed for tucked elytra (*tke*) in LO and bent elytral tips in CS purely on the basis of the expression of the elytra in some pupae. In these the elytral tips are folded under their more proximal portions resembling truncated elytra in both species of beetles.

The elongated elytra mutants in LO, CS, and CF may or may not be caused by homologous genes. As the name indicates, the elytra extend beyond the tip of the abdomen (an event which is, in itself, unusual), in CS and CF by the equivalent of the length of the posterior apparent abdominal segment, but in LO only slightly. However, *ele* in CF is identical to *ele* in CS.

The warped elytra and droopy elytra conditions in LO, CS, and CF are probably phenodeviants in all these species. Homology, therefore, would be difficult to establish.

The bent tibia (*btt*) mutation is incompletely penetrant in all species of *Tribolium* listed in the table. The phenotype of this gene in the pupa or in the adult is identical in strongly expressed beetles: the tibiae, particularly of the hind legs, acquire a double bend. Establishment of homology depends on linkage relationships, since other genes, affecting other parts of the legs, produce individuals which have an expression of the tibia similar to that produced by *btt*. At present it is not clear whether beetles homozygous for the other genes affecting other podomeres were also homozygous for *btt*.

Sternites incomplete (*sti*) has been found in *T. anaphe*, CS, and CF. While the *sti* condition sometimes appears in association with some of the split mutants, in all these cases the elytra appear normal. Hence, the *sti* condition may be due to homologous genes.

The creased abdominal sternites (*cas*) condition in LO and CF was found linked with pearl (see above). It is quite likely that *cas* in CS and in *T. destructor* may also be located in the pearl linkage group.

Melanotic stink glands in CS and CF is also, highly probably, an example of homologous genes in action. In both species the contents in the reservoirs of the odoriferous glands, which normally is a yellowish liquid consisting of quinones, become modified into a black solid mass visible through the exoskeleton. This mass is a polymeric material of high molecular weight which in CF has proved to be unanalyzable by spectroscopic analysis. This material has not been analyzed in CS.

Thus, it is evident from the preceding that the cases of mutants in species in the same genus or species in different genera exhibiting similar phenotypes are by no means rare. Future linkage studies will be able to discriminate between genes which exhibit parallel phenotypes but are not homologous, and those which exhibit the same, or nearly the same, phenotype but are homologous.

X. Possible Homologous Genes in Higher Categories of Coleoptera

The number of extant mutations known outside of tenebrionids is small. Dyte and Blackman (1961) have reviewed the situation in regard to eye color genes, most of which have been found in stocks maintained in the laboratory for many years. For the Dermestidae they cite Philip (1940) who described a white-eyed mutant in *Dermestes maculatus* comparable to pearl in *Tribolium castaneum*, and Reynolds and Sylvester (1961) who worked out the genetics of pearl in *Trogoderma granarium*. Both of these behaved as autosomal recessives, and Dyte and Blackman (1961) found that the larval ocelli and the frontal ocelli and compound eyes of the adult in the last species were unpigmented. A pearl mutant, also behaving as an autosomal recessive, has been found at the Pest Infestation Laboratory, Slough, Bucks., England, in *Carpophilus dimidiatus* (Nitidulidae) in which, again, the ocelli in the larva and the compound eye in the adult were pigmentless and pearl-like. Amos and Scott (1965) have determined that pearl is governed by a single recessive autosomal gene.

To their list of autosomal eye color mutants may be added a white mutation reported by Breitenbecher (1926) for *Bruchus* (*Callosobruchus*) *quadrimaculatus* (Bruchidae) and the yellow, milky, pearl, and apricot mutants described by Bartlett (1964a,b, 1965) for the curculionid *Anthonomus grandis*. (See also Chapter V for other additions.)

Finally, in *Cryptolestes turcicus* (Cucujidae) a reddish eye mutation has been found (Dyte and Blackman, 1961). This mutation appears to be sex-linked (Dyte *et al.*, 1965) and it would represent the first sex-linked eye color mutation outside the Tenebrionidae.

Aside from the "apterous" mutation reported by Breitenbecher (1925) for the bruchid *Bruchus* (*Callosobruchus*) *quadrimaculatus*, the mutant "*Divergens*" described for the coccinellid *Epilachna chrysomelina* by Timofeeff-Ressovsky (1935) and the mutants "bashful" and gnarled noted by Bartlett (1964a, 1965) for the curculionid *Anthonomus grandis*, no other body abnormalities attributable to genetic effects have been reported for beetles in other families. Fortunately, however, many entomologists specializing in Coleoptera often report teratologies. It is clear from the extensive review of teratological literature by Balazuc (1948), Mocquerys (1880), etc., that some of the cases reported as teratologies probably had genetic basis. A few of these will be cited here, since their phenotypes largely resemble those of mutants in *Tribolium*, and they probably represent instances of homology at the family level.

Von Heyden (1881) described an anomalous *Aromia moschata* (Cerambycidae) with elytra, legs, and antennae atrophied. Its phenotype largely resembles the miniature appendaged condition in *Tribolium*.

Von Lengerken (1933) reported a teratology in the cerambycid *Trachyderes succinctus* whose antennae were modified into legs resembling the homeotic mutant antennapedia in *T. castaneum*. The presence of heteromorphic organs has also been recorded in other beetles: in the cerambycid *Strangalia quadrifasciata* the left antenna terminated in a tarsus (Przibram, 1919); in *Telephorus* (*Cantharis*) *fuscus* (*Cantharidae*) and *Carabus festivus* (Carabidae) one of the antennae was bifurcated, and one of the branches appeared tarsuslike (Przibram, 1919). *Carabus violaceus* var. *fulgens* Charp. had the eleventh segment of the antenna in the form of an onychium (Gadeau de Kerville, 1923). In view of the fact that weaker expressions of antennapedia are possible, these abnormalities may have had some genetic basis.

Balazuc (1948) figures a specimen of *Lagria hirta* (Lagriidae) with processes arising from the femurs of all legs and from the antennae. It has already been pointed out that in *T. castaneum* the *sta* (spikes on trochanters and antennae) gene does very much the same thing, and in many cases produces triple-branched antennae. (The triple branching of the antenna, however, may result from the action of more than one gene, as evidenced from the fact that Dawson (1962a) was able to selectively increase the incidence of the phenodeviant called *bra*, branched antenna, or it may result from a developmental accident.)

The teratological literature abounds with cases of single specimens found in taxonomic collections with various parts of the legs deformed or branched, elytra blistered or divergent, prothorax deformed or subdivided into two or more components. In *Tribolium* some of these abnormalities proved to be solely due to accidents of development (Sokoloff, 1960b) while other, similar, deformities proved heritable. Hence, it is likely that many of these examples of teratologies must have been under genetic control, but it would be difficult to prove since many of these families of beetles are difficult to rear in the laboratory. There are, however, beetles in the families Curculionidae, Cucujidae, Dermestidae, Nitidulidae, etc., which are pests and easily reared in the laboratory. The study of these organisms from the standpoint of comparative genetics would yield interesting results.

It may be pointed out that Crowson (1960) has concluded that polyphagan beetles must have originated in the Jurassic. So far the evidence appears very strong in favor of homologous genes at the subfamily level. It is felt that Coleoptera may also provide evidence for homology of genes at least at the family level.

XI. Genetics of Populations
of Tribolium Species

A detailed review of the findings of investigations utilizing *Tribolium* as the test organism in various problems in population genetics is beyond the intended scope of this paper, but some idea of the areas of research in which flour beetles are used may be gained from the following (arranged more or less in chronological order).

Kollros (1944) was the first investigator to use *Tribolium* in studies of natural selection. She introduced the mutant pearl (p) in *Tribolium castaneum* at different initial frequencies. The generations were allowed to overlap in these populations. She found that this gene was selected out, but the reason for its elimination was not clear, since in many characteristics pearl equaled or exceeded wild type. The performance of $+/+$, $+/p$, and p/p for various attributes tested can be summarized as follows:

1. The rate of development of the egg stage was in the relations: $+/p > p/p > +/+$
2. Size: $+/p > +/+ > p/p$
3. Fecundity (total eggs laid): $+/p > +/+ > p/p$
4. Fertility: $+/p > p/p > +/+$
5. Longevity: male $p/p > +/+ > +/p$; female $+/p = p/p > +/+$; total $p/p > +/p > +/+$.
6. Larval survival: crowded $+/p > p/p > +/+$; uncrowded, same results.
7. Cannibalism: male $+/p > +/+ > p/p$; female $+/p \gg p/p > +/+$
Some populations gave a higher frequency of pearl than of normal and it was observed that in these the size of the populations was larger. No advantage to the heterozygotes could be detected.

Park (1948, 1954, 1957) has intensively investigated interspecies competition between *Tribolium confusum* (to be referred to in the following as CF) and *T. castaneum* (CS). His results may be summarized as follows: under environmental conditions of 24°C and 30% relative humidity CF invariably eliminated CS, whereas at 34°C and 70% relative humidity CS always eliminated CF. At intermediate conditions (29°C and 70% relative humidity) one species was found to be the victor, sooner or later, but the results were said to be indeterminate since it was not possible to predict, *a priori*, which species would win in a given experimental vial. To test whether competitive ability has any genetic basis, Park and Lloyd (1955) selected, from CS-CF competition vials, the occasional CF winner, and had it compete again with the original CS strain. The results of these experiments were the same as the previous ones: CF again was eliminated by CS in most cultures. These experiments are reviewed, and a statistical model developed, by Neyman *et al.* (1955). According to this mathematical model the outcome of competition is essentially a stochastic process, i.e., chance events determine which particular species (CS or CF) would win in an experimental vial.

Lerner and Ho (1961), suspecting that genetic variation in competitive ability must exist in *Tribolium*, repeated Park and Lloyd's experiments, using synthetic strains of CS and CF and an identical experimental design to that used by Park and Lloyd except that ten pairs (instead of two pairs) of founders were initially introduced. In these experiments CS eliminated CF in all vials. The experiment was then repeated using two inbred strains of CS and two of CF, propagated by brother-sister matings for thirteen generations, and hybrid strains resulting from crosses between the inbred strains. In order to reduce the time needed to complete an experiment, Lerner and Ho modified the technique. That used by Park entailed starting the populations with two pairs of beetles of each species, the return of all the live stages to the vial every month when the medium was renewed and the beetles censused. In Lerner and Ho's experiments the founding populations consisted of ten pairs of beetles of each species, and the adults were discarded every 30 days after censusing: the medium was renewed and the competition vials continued with the preimaginal population. This technique gave the same outcome of competition but clear-cut results were obtainable in about 5–8 months after initiation of the experiment instead of the 18–22 months required using Park's design.

In these experiments there were three types of results: (1) determinate, with CS the winner; (2) determinate, with CF the winner; and (3) indeterminate, with CS superior on the average but with an occasional CF win. Lerner and Ho's results indicated clearly that if the genotype of the culture for competing ability is known, the outcome of competition can be determined. Lerner and Dempster (1962) suggested that much of the indeterminacy observed in Park's experiments may be a reflection of random selection of the genotypes of founder populations. Unpublished data using the same strains used by Lerner and Ho and demonstrating genetic differences in competitive ability but using Park's method of handling the cultures suggest that the differences are not due to a difference in experimental design. Independent evidence for the relationship between genetic variation and indeterminacy was provided by additional data by Dawson and Lerner (1962). Park *et al.* (1964) have recently published additional data on the performance of competition of CS and CF utilizing inbred strains. Their results suggest that the use of inbred strains gives deterministic outcome of competition conforming with the findings of Lerner and Ho (1961), Lerner and Dempster (1962), and Dawson and Lerner (1962).

McDonald and Peer (1960) used *Tribolium confusum* to follow natural selection of Striped, *St* (an autosomal dominant with recessive lethal effects) and the semidominant black (*b*) in population cages. Changes in *St* were followed from the female segment of the population since *St* males have greatly reduced viability and are sterile. The initial frequency of *St* (which exhibits, in females, a 20% lowering in viability) was 0.95 and of wild-type females 0.05. *St* declined rapidly when single discrete generations were followed, so that by the end of five generations the frequency of *St* females was about 0.01. The curve produced by the decline in each generation was very close to that given for an adaptive value of 0.4. In population cages where generations were allowed to overlap, a gradual decline to about 20% *St* was observed during the first 38 weeks. Thereafter the decline of *St* was more rapid, becoming extinct by the fifty-second week. Mortality and natality data could be used to interpret more adequately the observed changes in frequency of *St*.

In the same paper, McDonald and Peer followed the changes in frequency of the semidominant gene black, *b*, in populations of *T. confusum* where generations were allowed to overlap. From an initial frequency of 0.5, the *b* gene declined to a frequency of about 30% in the two replicate cages after a period of observation of 125 weeks. Assuming *b/+* to have a fitness of 1.00 and an interval of 18 weeks per generation, the adaptive values of *+/+* and *b/b* were 1.30 and 0.95, respectively.

McDonald and Peer (1961a) investigated the fitness of the *T. confusum* mutant split (*sp*) in populations allowed to produce overlapping generations. The population cages were started with four wild-type and 462 *sp* adults. By the end of 10 weeks the cages contained 30% or less *sp* adults, and by the end of 26 weeks the *sp* gene had beeen eliminated. The loss in fitness was considered to be primarily due to a lowering in viability and productivity of mutant females, since they produce only about one-sixth as many progeny as wild-type females; *sp* males are as fertile as normal males, but each type of male is more productive with its own type of female. On the other hand, mutant males copulate less readily and effectively than wild-type males, whereas mutant females accept wild-type males as readily as mutant males.

Bray *et al.* (1962) used *Tribolium castaneum* in selection experiments designed to test the importance of genotype-environment interaction with reference to control populations. They tested fifteen methods of maintaining control populations over eight generations in regard to their ability to establish the level of environment in respect to the base population and to separate environmental and genetic effects in lines selected in two directions. All the lines tested were grown in one environment for two generations and for two more in another; the beetles were selected for large and small pupal weight in a dry (40% relative humidity) and a wet (70% relative humidity) environment. Preliminary studies had shown that the control (foundation stock) populations produced pupae weighing 10% less in the dry than in the wet environment. Selection in opposite directions in regard to pupal weight gave a clearly symmetrical response. The response of inbred lines derived from different stocks was different in the two environments. Evidence for genetic drift was obtained for the mass-mated lines, but not for other methods of maintaining the original population, presumably because of the large numbers of families tested. It was noted that greater progress in selection was obtained in the dry than in the wet environment. Repeated and relaxed methods of maintaining controls were more effective in showing how environmental changes affected later generations of the selected lines.

Schlager (1963) combined ecological and genetic methods in studying sooty (*s*), a semidominant body color gene in *T. castaneum*. Five types of populations were established, containing the following initial proportions of *s*: 1.00 *s*, 0.75 *s*, 0.25 *s*, and 0.0 *s*, introduced in 100 pairs of adults in 40 g of flour. The growth rate, maximum number of individuals, asymptotic number of adults, death rate of adults and number of adults replaced, were highest in the 0.75 *s* and lowest in the 0.25 *s* populations. Greater deviations in the differences between adult zygotic frequencies of two successive census periods were observed in the 0.75 *s* and 0.50 *s* than in the 0.25 *s* populations. These periods remained fairly constant for the remainder of the study and apparently did not result from any consistent deviation in random mating. The fate of *s* was different in various populations: it declined rapidly at first and then more smoothly in the 0.75 *s* populations; while in the 0.50 *s* it declined gradually during the entire course of the experiment. In the 0.25 *s* populations a marked increase occurred, after which *s* fluctuated around an equilibrium value of $q = 0.33$. A reasonably good fit for several sections of the population cage study was given for adaptive values of 1.05, 1.00, and 0.95 for $+/+$, $+/s$, and s/s, respectively. However, adaptive values changed, possibly due to a change in the nature of the selection process, in the 0.75 *s* and 0.25 *s* populations. Assays of fecundity and fertility for the three genotypes accounted in part for the changes in gene frequency observed. Studies of developmental rates showed that s/s develops more slowly than $+/+$ and $+/s$ in densities of 20 animals/g medium. At densities of 50 animals/g medium $+/s$ developed faster than either homozygote, and there was a slight increase in mortality of s/s. Schlager concludes that since the relative performance of the three genotypes changes as density increases, the observed changes in gene frequency are dependent on density. The magnitude of these changes is also dependent on the frequency of *s* in the population.

Sokal and Huber (1963) utilized the sooty mutant to determine the competitive ability of the various genotypes at varying densities and gene frequencies. Eggs, collected from egg farms, were placed in rearing vials containing 8 g of medium at frequencies of *s* equivalent to 0.0 q_s, 0.25 q_s, 0.75 q_s and 1.0 q_s. For each gene frequency, four densities were set up: 5/g, 20/g, 50/g, and 100/g. $+/+$ did better than $+/s$ and s/s at the lower densities for gene frequencies 0.25 q_s and 0.5 q_s. At gene frequency 0.75 q_s $+/s$ appears to be favored over $+/+$ and s/s at the lower densities. The highest density depressed all survival values somewhat, but here $+/+$ and s/s did better than $+/s$. The values for the 50/g density are intermediate between those at the two lower densities and those at density 100/g. These results indicate that $+/s$ improves in performance as the gene frequency of *s* increases. Assays included in this study were dry weights of adults which proved to be highest at 20/g. The three genotypes in mixed culture weighed as follows: $s/s \gg +/+ > +/s$, but the gene frequency in the cultures had marked effect on weights of $+/+$ and $+/s$ individuals. Weight was not simply a function of the gene frequency of *s* in the culture but of the two genotypes carrying the *s* allele. Heterozygotes were found to be intolerant to crowding, especially with their own genotype, and developed much faster than s/s.

Sokal and Karten (1964) investigated competition among genotypes using the black (*b*) mutant. Gene frequencies in pure and mixed cultures were the same as those indicated above for the sooty mutant. The densities per vial were also the same. Increase in density caused a decrease of adult survivorship in $+/b$ and b/b but not of $+/+$. Genetic facilitation (especially by the heterozygotes) was experienced by all three genotypes in mixed cultures, the degree of facilitation for each genotype differing, depending on the gene frequency and the density conditions. In almost all combinations of gene frequency and density, the $+/b$ have the highest survivorship. At lower densities b/b is better adapted than $+/+$, while at higher densities and with increasing gene frequencies of b, $+/+$ appears to be better adapted. Adaptive values of the various genotypes were found to be gene-frequency dependent. As the gene frequency of b increases, the emergence of $+/b$ and b/b decreases, but that of $+/+$ increases. Mean dry weights of adults are in the relation $20/g > 50/g \gg 100/g$. The weights of the three genotypes were in the order $+/b > +/+ > b/b$ but no gene-frequency dependence effects on the weights were observed. Developmental period of the genotypes was in the order $+/+ > b/b \ll +/b$, and in relation to density $100/g \ll 5/g > 20/g > 50/g$, but in pure cultures these relationships change somewhat at the higher densities. On the whole, changes in gene frequency have no effect on developmental period, but some effects could be observed at certain densities.

When the homozygous b/b genotype was rare (1%) or frequent (81%) genetic facilitation was present, but when $+/+$ was rare it showed a marked increase in adult survivorship. The weights of $+/+$ are not changed when they are frequent. On the other hand, b/b, when rare, have lower weights, but when frequent they weigh more at all densities. Sokal and Karten conclude that as gene facilitation and gene-frequency-dependent adaptive values must be widespread phenomena (since they have been observed in *Drosophila* and other Diptera), they may complicate simpler models of selection and appreciably modify conclusions based upon them.

Two recent studies have been made concerning the genetics of developmental rate in *Tribolium*. Englert (1964) selected for early and late pupation in *Tribolium castaneum*. His lines, originating from a heterogeneous random mated foundation population, were reared at 33°C and 70% relative humidity for six generations of selection. A parallel study attempted to detect whether 13-day larval weight and pupal weight were affected by selection for pupation time. Selection for the latter gave an asymmetrical response: development was speeded up by almost 2 days in the lines selected for fast, and delayed by almost 5 days when selected for slow pupation time as contrasted with the controls. Realized heritabilities were estimated as 0.043 for early and 0.262 for late pupation. Phenotypic variance increased in late pupation lines and decreased in early pupation lines, while that of the controls fluctuated largely around the original estimates. Pupation time was highly correlated in an inverse manner with 13-day larval weight, values of —.760 and —.863 being obtained for early and late lines, respectively. Selection resulted in a decline in reproductive fitness in the late pupation lines but may have resulted in an increase in the early pupation lines. The decline was attributed to both a lowering in fecundity and hatchability. Hatching time, measured in the fourth generation, was increased by 1 hour in the late pupation lines but not affected in the early pupation lines.

Dawson (1964b) obtained response to selection for fast and slow development in CF. The total developmental time (egg to adult) in our synthetic CS and CF (at 29°C and 70% relative humidity) is about 31 days. By selection the CS fast developed in about 26 days and the CF fast in about 29 days. The CS slow emerged as adults in about 43 days. Analysis of developmental rates of inbred lines and hybrids obtained therefrom showed that aside from some evidence of maternal effects in some crosses, all of the hybrids showed heterosis for fast development and a reduction in variability.

On the basis of these preliminary findings Dawson proposed a testable model for the relationship between developmental rate and competitive ability with two assumptions: (1) the synthetic populations are at an adaptive optimum with respect to developmental rate, and (2) developmental rate is an important component of competitive ability. This model predicts that competitive ability would be decreased in selected lines. To test this model, lines selected for slow and fast development in CS, and fast development in CF were used in competition studies. The detailed analysis of all the various axillary studies cannot be cited here, but in summary all selected lines performed worse than the synthetic strains from which they were derived, leading to the same conclusion of Park *et al.* (1961), namely that "competitive ability is the sum total of many factors which interact in exceedingly complex ways." One strain of CS selected for fast development performed poorly in early selected generations but better than the standard CS strain in later generations. In general, Dawson's results of the competition experiments gave convincing evidence to support the postulated model for the relationship between competitive ability and developmental rate: "Artificial selection for developmental rate moved the synthetic populations from their adaptive peaks leading to decreased competitive ability. As competition proceeded, developmental rate in the CS-fast selected strain returned toward the original adaptive level roughly proportionally to the speed with which CF was eliminated" (Dawson, 1964b, p. 77).

Dawson concludes that "the CF synthetic strain possesses relatively greater homeostatic properties than the corresponding CS strain," developing this concept in terms of the natural conditions under which the two species are found and the outcome of competition between synthetic strains.

Crenshaw and Lerner (1964) have assayed productivity of strains inbred by brother-sister matings after eight to ten generations. For the strains investigated, significant differences in means and variances in length of productivity period were found for *T. castaneum* (CS) but not for *T. confusum* (CF). For total 120-day productivity, significant intraspecies, between-strain variation in means and variances was found for both species. Per-day productivity for a 40-day period revealed significant strain differences in CF, but interstrain differences in variance could not be established. On the other hand, in CS, significant interstrain variation in means and variances was present. Sibling strains of the one CS pair tested showed significant differences in means and variances in regard to productivity period and per-day productivity. In CF two sibling strains of one pair showed similarities in all traits compared; while in another pair slight but significant differences could be established only for 120-day productivity. Crenshaw and Lerner conclude that under competition conditions CS appears to perform better and CF relatively less well than would be expected on the basis of the productivity studies.

Sokoloff *et al.* (1965) have studied productivity of four normal laboratory strains (Chicago, McGill, Texas, and Virginia) and five mutant strains [four derived from the Chicago strain (Chicago black, paddle, pearl, and jet) and one derived from McGill (McGill black)] under optimal conditions (daily transfer of the imagoes to fresh medium) and suboptimal conditions (transfer to fresh medium each 20 days). Productivity under optimal was 20–40 times greater than productivity under suboptimal conditions. In general, mutant strains were less productive than the wild-type strains from which they were derived. Productivity was not influenced by age of female during the 80-day period covered by the study.

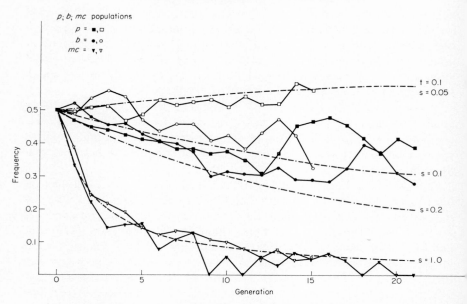

Fɪɢ. 4. Gene frequencies in two replicate populations initiated with 20 pairs of
$p/+$; $b/+$; $mc/+$. The dashed curves are derived from various theoretical selective
values.

Their results emphasize the fact that productivity of a strain cannot
be specified generally but only for a given set of conditions, and must
be empirically determined for each set of conditions.

Sokoloff (1964i) and Sokal and Sonleitner (1964), independently, have
investigated the possibility of applying the technique devised by Lerner
and Ho (1961) in studies of natural selection involving various genes
in populations. As has been previously pointed out, Lerner and Ho,
in their competition studies, censused and discarded the adults found
at intervals of a month, at which time the medium was renewed and
the populations continued with the juvenile stages. Sokal and Sonleitner
(1964) used the black (b) mutation in *T. castaneum*, starting their populations with heterozygotes. Throughout their study, the parents were allowed to lay eggs for 3 days, after which the adults were discarded.
The eggs were allowed to complete their development, and all the adults,
after their respective phenotypes had been recorded, introduced into bot-
tles containing specified amounts of medium (40 g). After a 3-day
laying period the parents were again discarded. This procedure, followed

for each generation, resulted in populations of the order of several thousand, and the period for completion of development of all the eggs into adults lengthened from an initial 30 days to over 40 days. The gene frequency under this scheme increased gradually to 0.8.

Sokoloff (1964i) used the same black mutation as well as two other genes in separate linkage groups (pearl, p, and microcephalic, mc). Replicate populations were started with adults heterozygous for one gene, two genes, or all three in two different media, but only the results in whole wheat flour will be mentioned. At monthly intervals the adults were censused and discarded, and the preimaginal stages returned to containers with 50 gm of fresh medium. By this scheme populations averaged between 400 and 645 adults. The changes in gene frequencies of two replicates where all three genes were involved are shown in Fig. 4. It is clear that in these populations the fate of each gene is independent of the fate of the other genes when all three genes are present in the same population. These results are not different from those in populations where a single gene has been introduced in heterozygous form. Thus, it can be concluded that the Lerner and Ho method can be utilized safely in such studies of natural selection. Obviously, the advantages of such a procedure are that twelve artificially discrete generations can be followed per year, and the interpretation of the data is greatly simplified.

XII. Contributions of Tribolium Genetics to the Problem of Morphogenesis of the Antenna in the Coleoptera

The mutant material described elsewhere in this review is of interest in one other area, that concerned with morphogenesis of the antenna in the Coleoptera.

It is well known that taxonomists rely on the stability of the phenotype of numerous morphological characters in building up their taxonomic keys. This stability, ultimately, is a reflection of the hereditary material and the faithful transcription of the encoded message in the DNA molecule to manufacture enzymes and proteins. If the beetle's DNA is normal a normal character is produced, while if the DNA is modified, a mutant form of the character is formed. *Tribolium* may or may not be suitable for more refined studies giving an insight into the biochemical changes between gene and character, but for the present purposes these details are not necessary.

The antennae and/or legs of beetles are suitable end-product organs in which the morphogenetic pattern may be deduced, since these appendages generally remain intact in teneral normal and mutant adults. (The only exceptions are those beetles which, during critical stages in their development, are exposed to high concentrations of quinones or other gases. These beetles may lose part or the whole antenna or other appendages during metamorphosis.)

In tenebrionids the antennae consist of eleven segments. There are certain similarities and differences in form in different species. For example, *Tribolium, Latheticus,* and other genera possess an antenna consisting of two basal segments (pedicel and scape) and a flagellum. The latter is subdivided into a club and a funicle. The club may consist of five segments in different genera (*Tribolium, Latheticus*), but it may consist of a variable number in species within the same genus (three in *T. castaneum,* and five in *T. confusum*). The shape of the funicle and club may differ: in *Latheticus* these portions of the flagellum are roughly spindle-shaped owing to a terminal tapering of the club segments (Plate

16, A); in *T. confusum* the funicle consists of four segments about equivalent in size, but the five segments constituting the club gradually expand toward the distal end (Plate 11, E). Finally, in *T. castaneum* there are six funicular segments. The one next to the club is considerably larger, but not as large as the first club segment. All segments constituting the club in the normal beetle are about equal in size (Plate 6, A).

Since the scape and pedicel are nearly identical in shape in all three species and since they are not affected in most antennal mutants, one could suppose that these antennameres form independently of the flagellar segments. Further, since the flagellum has two distinct regions (funicle and club), it could be supposed that these regions form under the influence of different genes. Furthermore, since the antennae differ so much in phenotype, one could invoke different systems in effect in the formation of these organs.

The data collected on antennal mutants in *Tribolium* and *Latheticus* (chiefly by Sokoloff) indicate that *all* the antennameres may come under the influence of certain genes (the pedicel and scape are not immune to their effect) and that some genes affect some segments more often than other segments. As can be seen from Tables 3–12, 14–16, and 17, the antennal mutants exhibit a fairly regular pattern regarding the annuli or antennameres involved in the various fused blocks. These data have been reworked to show the frequencies of adjacent segments exhibiting fusion, and they are summarized in Table 22.

The data in the Table show that the paddle (*pd*) and serrate (*ser*) sex-linked genes behave differently in the male than in the female. In the latter, the effect of the two genes is confined primarily to the club, and when funicular segments are affected, usually a moderate to long gap between groups of affected antennameres is filled by normal segments. In paddle males, despite much stronger fusions, it is still possible to identify each of the segments entering into the fusions, and it becomes evident that segments 4–11 are involved. In this sex there are no intercalations of normal segments between fused blocks, and it is of interest that the frequency of involvement of the various antennameres drops toward the proximal end of the antenna, and segments 1–3 (from proximal to distal) are free of fusions. The effect of *ser*, while limited essentially to the club in females, in the male extends more proximally than the effect of *pd*, since segments 3–4 are fused. Although not included in the Table owing to the small numbers of beetles reported, the *ser* gene in the male may affect the pedicel with the result that the latter fuses with antennal segment 3 (the first segment of the funicle). From the

TABLE 22

Frequency of Fusion of Antennameres in Various Mutants in *Tribolium castaneum* (CS), *Tribolium confusum* (CF), and *Latheticus oryzae* (LO).

Species	Mutant	Annuli fused													Total fusions	N*
		None	1–2	2–3	3–4	4–5	5–6	6–7	7–8	8–9	9–10	10–11	11–12	12–13		
CS	pd ♂	—	—	—	—	17	29	50	99	143	193	119	—	—	650	98
	pd ♀	4	—	—	—	—	—	—	11	—	183	10	—	—	204	101
	ser ♂	—	—	—	3	17	3	9	1	3	6	9	—	—	51	10
	ser ♀	17	—	—	—	2	—	—	—	—	3	7	—	—	12	10
	fas-1	7	—	1	3	129	9	9	—	3	183	—	—	—	341	116
	fas-2	7	—	1	—	17	103	—	1	1	12	—	—	—	154	65
	fas-3	49	—	—	1	74	7	70	—	—	—	—	—	—	—	—
	fas-3a	—	—	—	75	77	—	78	—	43	36	42	—	—	358	40
	Fas-4	—	—	—	82	49	111	47	58	—	62	—	—	—	304	50
	Fas-5	—	—	—	6	86	14	—	—	—	1	1	—	—	213	50
	aps	8	—	—	1	37	4	17	64	26	91	75	16	2	341	57
	pec	4	—	4	14	36	26	15	16	4	9	1	—	—	129	28
	Spa	2	—	—	2	6	15	17	22	3	25	35	—	—	127	20
	Sa-2	—	—	118	118	118	113	108	100	—	—	—	—	—	675	60
CF	fas-1	—	—	—	—	—	—	—	52	9	1	1	—	—	62	40
	fas-2	—	—	—	30	—	30	—	—	12	7	—	—	—	80	20
	st1	2	—	2	1	7	7	1	28	—	35	22	—	—	105	35
LO	fas-1	—	—	—	—	23	5	11	8	67	2	4	—	—	120	69

* Number of beetles observed.

small sample of *ser* beetles reported it is evident that the club segments and segments 4–5 of the funicle are the ones most often affected. Noteworthy, in comparing the two sexes, is the fact that when funicular segment 8 is fused with club segment 9, the number of annuli involved in fusions toward the proximal end increases. If segments 8–9 are not fused, the more proximal segments may be free of fusions.

Turning now to the autosomal mutants in *T. castaneum*, it is evident that the *fas-1* gene affects segments 4–5 and 9–10; *fas-3* affects 4–5 and 6–7, and *Fas-4* affects primarily 3–4 and 5–6 of the funicle and 9–10 of the club. In all these mutants, abnormal portions of the antenna are separated by larger or smaller numbers of intervening antennameres which exhibits no fusions. Interestingly enough, *fas-3a*, a stronger allele of *fas-3*, affects more proximal segments of the funicle and more distal segments of the club. The genes *Fas-5* and *Sa-2* have their effect primarily on the funicle. The allele of antennapedia, *ap^s*, is also of interest because segments 8–9 are again involved in an antennal mutant which produces segments in excess of the normal, and yet the new segments respond to the aberrant code resulting in fused segments.

The *fas-1* gene in *T. confusum* operates within the five-segmented club, whereas *fas-2* and *stl* affect the club *and* the four-segmented funicle. The effect of the latter, in addition, involves the pedicel in a fusion of this segment with the first antennamere of the funicle. Finally, in *Latheticus*, the *fas-1* gene exerts its effect on the funicle and the club.

It may be stressed that in *T. castaneum* the pedicel may fuse with the funicular segments. This is evident from the tabulated information for *fas-2*, *pec*, and *Sa-2*, and, although not apparent from the table, this can also be observed in the *ser* and *Spa* mutations. The *stl* mutation also involves the pedicel. More rarely, the scape may be fused with the pedicel in such mutations as Fused tarsi and antennae and in the Short antenna alleles (*Fta*, *Sa*, and *sa*, respectively).

It is evident from this survey and Table 22 that:

1. There is considerable regularity in the fusions produced by a given gene.

2. There are similarities in the distribution of fusions produced by some genes, and differences among the distributions produced by other genes.

3. In all species the antennal mutants exhibit fusions, predominantly in the distal half of the antenna. The proximal half of the antenna has considerably fewer fusions.

4. In *T. castaneum* the involvement of funicular segment 8 with seg-

ment 9 of the club segments appears to result in an increased distribution of fusions of more proximal segments. This is true in the males of sex-linked genes (*pd, ser*) and in both sexes of autosomal genes. When segments 8–9 are fused, there is a greater chance that segments 4–5 and 6–7 will be found fused.

5. Neither the pedicel nor the scape are immune from producing fusions with more distal antennameres, although these proximal annuli are frequently free from fusions.

6. Some genes (*pd* in the female, *ser* in the female, *fas-1, fas-2, fas-3, ap^s, pec,* and *Spa* in *T. castaneum,* and *stl* in *T. confusum*) may produce one normal antenna and the other in the same individual will exhibit fusions. No ready explanation can be offered. In connection with these exceptions it may be pointed out again that, at least in *T. confusum,* the orientation of two adjacent fused blocks (each block consisting of two segments) is different than that of two other blocks separated by a normal antennamere (Plate 15, B).

The most reasonable hypothesis to explain all the above phenomena is that at least some of the genes affecting the antennae begin to exert their effect at the distal end of this appendage. The longer the genes exert their effect, the greater the numbers of more proximal segments exhibiting fusions. This holds true for such genes as *pd, ser, fas-2, fas-3a, Fas-4, ap^s, pec,* and *Spa* in *T. castaneum, stl* in *T. confusum,* and *fas-1* in *L. oryzae.* In the case of some genes (*fas-1, fas-3, fas-3a, Fas-4, Fas-5, ap^s,* and *Spa*) blocks of affected segments are separated by one or more normal segments. This is interpreted to mean that the effect of these genes, again beginning at the distal end of the antenna, is not constant, but that the effect of the gene is temporarily shut off allowing some segments to form free of fusions.

Clearly, timing is an important factor in the development of the antenna as it is in the rest of the developing organism. By way of illustration let us consider the *Sa-2* gene. This gene does not affect the scape or the club, but it does affect the pedicel and the funicle. If it is correct that the effect of most genes starts at the club end, the lack of fusions in this portion of the antennal appendage indicates the *Sa-2* has a delay in its action, but once it begins its effect it is constant and long-lasting and results in the fusion of the pedicel with the first funicular segment.

Dawson (personal communication, 1965) on the basis of his observations on the *ser* gene and the writer's recorded observations on the distribution of fusions produced by the *Fta, fas-1, fas-2,* and *fas-3* genes as well as some data on the *pd* mutation suggests an alternative hypothesis. This hypothesis is based on the 1917 work of Tillyard who studied the development of the adult antenna in the hemimetabolous dragonfly. According to Tillyard (see diagram in Imms, 1940) the antenna in this insect begins as a three-segmented structure. The two basal segments represent the scape and the pedicel and the much larger distal one the flagellum. The first two segments, once formed, do not subdivide while the third undergoes a series of divisions. The first division gives rise to a basal and a distal mass. The latter remains quiescent until the last nymphal instar when it subdivides into the last two terminal segments of the adult antenna. By this time the basal mass subdivides forming two components, and the most basal component subdivides again. Thus, from the intermediate mass arise three antennal segments which lie between the two distal and the two proximal antennal segments.

Using this model as a pattern, Dawson suggests that the adult antenna in *Tribolium* first appears to consist of three components: the basal two give rise to the scape and the pedicel, while the distal one will give rise to the flagellum. The first division of this component results in two masses; one, the distal one, will give rise to the club segments, and the proximal one to the funicle. The latter forms in the following sequence: the masses resulting from the first division will give rise to segments 3–4 and 5–8. A further subdivision of the 5–8 mass results in two components, one which will give rise to segments 5–6 and the other to segments 7–8. By this time the presumptive club mass has undergone two divisions. The first division gives rise to segment 11 and a mass which, by further division, will give rise to segments 9 and 10.

In seizing upon Tillyard's model Dawson ignores completely the effect of various genes (including his own *ser* mutant) on the pedicel and the scape in *Tribolium*.

Admittedly, the test for the correct model for antennal morphogenesis in the Coleoptera may require the development of special techniques. However, it is of interest to note that Pohley (1959) in his studies of the influence of amputation on the development of the antenna in *Periplaneta americana* reports that the development of the flagellum of the normal antenna in this insect occurs during the mitotic period between two molting intervals, and proceeds in a basipetal direction. Thus, the evidence derived from the phenotype of antennal mutants suggests that a similar process is in operation in *Tribolium*.

Acknowledgments

The writer has benefited from the able technical assistance of the following individuals at various times during his research on genetics of flour beetles: Mesdames Alma Carle, Hilda Fennessey, and Anne Rennell at the W. H. Miner Agricultural Research Institute, Chazy, New York; Misses Claire Dryfuss and Ann C. Frankowski, Mrs. Barbara Sokoloff, and Messrs. Peter P. Hofman, Austin E. Kreutz, Brackett L. Scheffy, and Roger St. Hilaire at the Biological Laboratory, Cold Spring Harbor, New York; and Mesdames Mary Ackerman, Marcia Anderson, Louise Bielfelt, Barbara Daly, Merlene Hohmann, Marjorie Hoy, Louise F. Overton, Clara Stephenson, and Barbara Strong; Misses Arlene Adachi, Mellissa Cargill, Ann Hofer, Elizabeth Michener, and Yuri Uyeno; and Messrs. Robert Coleman, George Ferrell, Frank K. Ho, Nobuo Inouye, Richard Paige, Peter Rodriguez, Walter Spencer, and Roger S. St. Hilaire at the Department of Genetics, University of California, Berkeley, California. Mrs. Barbara Daly and Mr. Roger S. St. Hilaire are responsible for most of the illustrations. Dr. Howell V. Daly, Department of Entomology and Acarology, University of California, Berkeley, kindly prepared the phase-contrast photographs shown in Plates 18 and 19. Permission from Dr. A. Wilkes, Editor, *Canadian Journal of Genetics and Cytology* and Mrs. Jaques Cattell, Publisher, *The American Naturalist,* to reproduce figures from earlier publications (cited where appropriate) is gratefully acknowledged. A special expression of gratitude is directed to my colleague Dr. I. Michael Lerner, who has tacitly encouraged the continuation of the investigation of the basic genetics of flour beetles by allowing complete freedom to pursue whatever line of research interested me. Occasionally I have also benefited from the advice of various colleagues, in particular Dr. Theodosius Dobzhansky and Dr. Bruce Wallace. This investigation was supported, at different times, by grants G-4501, GB-1779, and GB-4411 from the National Science Foundation, and grants RG-7842 and GM-08942 from the U.S. Public Health Service.

Appendix

During the course of my literature search it became evident that Tower's work was not cited very extensively, even a few years after his work appeared in print. In the United States, this work is briefly cited in H. H. Newman's "Evolution, Genetics and Eugenics" (1932), and Richard Goldschmidt's "Physiological Genetics" (1938). The most extensive discussion of Tower's work appeared in Castle's "Genetics and Eugenics" (1925). He states (pp. 69–71):

> In this country, W. L. Tower (1896) has carried on extensive experiments upon potato beetles and related insects, in which variations in temperature and humidity of the environment have been followed by variations in pigmentation of the insects, similar to those observed by Fischer in the case of butterflies. Tower interprets his observations, as would Weismann, as showing, not inheritance of acquired characters but direct modification of the germ cells, independently of the soma. For, he claims to have obtained modification of the germ-plasm, which accordingly resulted in inherited variations where no parallel modifications of the body of the parent had occurred. Inheritance of an acquired character is accordingly excluded because no modification was acquired. His strongest evidence for this claim consists of cases in which the same parents were subjected to periods of heat or cold, alternating with periods of normal temperature, each being of several weeks duration. It was found that when a batch of eggs was produced in or immediately following a period of heat, characteristic color variations were likely to occur among the offspring which may be called heat variations and they proved hereditary. But when eggs were produced by the same parents at normal temperatures, no such variations occurred. Similar effects were obtained in cold periods, as contrasted with normal temperatures. While the bodies of the parents remain unaffected, the coloration of their offspring varied with conditions of temperature and moisture during the growth and fertilization of the eggs which produced those offspring. Tower therefore concludes that the germ-plasm was directly and permanently affected by variations in the environment during a particular sensitive growth period of the egg. This work is therefore no argument for the inheritance of acquired characters; nevertheless it is an argument for evolution directly guided by the environment which after all is the essence of Lamarckism. There are several reasons why we should accept Tower's conclusions with some reservation.
>
> 1. In the first place his experiments are not reported in sufficient detail to enable us to form a critical opinion as to their conclusiveness.
> 2. If the supposed temperature and moisture effects are due solely to those conditions, they should appear equally in all eggs subjected to the same conditions, but

this is not the case. Only certain individuals are modified. Since this is not s$\bullet$ it is evident that all the eggs were not alike at the outset for some were mo$\bullet$ sensitive than others to temperature and moisture changes in the environment, indeed these were the agencies which caused the changes observed. A good argume$\bullet$ could therefore be made for considering the temperature and moisture changes a$\bullet$ merely selective agencies exerted on a collection of germ-cells already inherent$\bullet$ variable in their potentialities. For Tower maintains that the variations once obtaine$\bullet$ are perfectly stable for an indefinite number of generations. His claim, therefor$\bullet$ that by direct action of the environment for a comparatively short period permane$\bullet$ changes in the germ-plasm were brought about. It would seem that if the germ-plas$\bullet$ is thus directly modifiable, the action ought to be reversible. Changes of environme$\bullet$ should *unmake* species as readily as they make them, yet such a result would scarce$\bullet$ harmonize with Tower's theory, or with the known stubborn and persistent natu$\bullet$ of heritable variations, when once they have arisen.

At the suggestion of Drs. Ernst W. Caspari and Curt Stern I hav$\bullet$ contacted the following individuals who were graduate students at th$\bullet$ time Tower did his work: Dr. L. C. Dunn, Nevis Biological Station$\bullet$ Irvington on Hudson, New York; Dr. A. H. Sturtevant, California Insti$\bullet$ tute of Technology, Pasadena, California; and Dr. Sewall Wright, Uni$\bullet$ versity of Wisconsin, Madison, Wisconsin. I inquired from them whethe$\bullet$ they had any personal knowledge of Towers' work, and if so, whethe$\bullet$ there was an impression among some of the older geneticists that Tower'$\bullet$ work was of doubtful validity and what were some of the criticisms. Be$\bullet$ cause the replies have historical interest, I have asked the writers of th$\bullet$ following letters for permission to quote them in part or *in toto:*

Columbia University in the city of New York
Nevis Biological Station
Irvington on Hudson, New York

June 14, 1965

Dr. A. Sokoloff
University of California
Department of Genetics
345 Mulford Hall
Berkeley 4, California

Dear Alec:

I am glad to hear of the progress of your review. I don't recall having suggested that Tower's work might have had fraudulent parts and certainly I have no personal knowledge to support such an accusation. I have not seen his papers for many years but recall being impressed with the thoroughness of the early observations on lack of effect of selection on spot pattern. The work on what we should now call induced mutation was, as I recall, subject to some difficulty in getting adequate controls as other early experiments of that kind.

Of course he should be cited but with some reservations as Castle did in the 3rd Edition of Genetics and Eugenics. I suppose he has not been cited in later works (although I have not checked Stubbe's 1938 Review of Mutation or Timofeeff's book of 1937) simply because later work on better known animals and plants left fewer questions unresolved than work which had not been repeated in an organism not much studied genetically.

With best wishes,

Sincerely yours,

L. C. Dunn

LCD:mh

California Institute of Technology
Pasadena, California 91109

Division of Biology

June 11, 1965

Dr. A. Sokoloff
Department of Genetics
345 Mulford Hall
University of California
Berkeley, California 94720

Dear Dr. Sokoloff:

I knew W. L. Tower slightly. He was a very interesting conversationalist. T. H. Morgan, who was not easily satisfied, once told me that he had once spent more time in a single long conversation with Tower than he had ever done with anyone else—and had enjoyed all of it.

Tower was very secretive about his work. He never showed people around his laboratory, and his own graduate students never knew what he was doing. One of them later told me that he had assisted Tower with the Leptinotarsa experiments, and that the experiments were always coded, so that he (the student) never knew what he was doing nor could he tell from the published accounts what he had been working with.

As he grew older Tower apparently became more and more secretive and withdrawn—perhaps because of the widespread skepticism that he encountered.

Perhaps the best way to state it is that the reliability of the data is questionable.

Sincerely yours,

A. H. Sturtevant

AHS: gc

The University of Wisconsin
College of Agriculture
Madison, Wisconsin 53706

Department of Genetics   June 28, 1965

Dr. A. Sokoloff
Department of Genetics
University of California
Berkeley 4, California

Dear Dr. Sokoloff:

The statement by Dunn on W. L. Tower, as recalled by Caspari certainly reflects a general impression after about 1911 of the quality of his work. Tower's experiments with Leptinotarsa had been looked upon as among the most important in genetics in this country. I recall reviewing his 1906 publication in 1911 as an undergraduate in a course on Evolution and being very much impressed by the combination of genetics, biochemistry and other fields that he brought to bear on the subject. A year later, however, I read a criticism of this paper by the Biochemist R. A. Gortner (who later wrote one of the most widely used texts on biochemistry). Tower had claimed that his experiments supported the view that the pigment was a diazo compound. Gortner stated that no one had ever maintained this, that Tower obviously had no familiarity with biochemistry, that his tests as far as they were intelligible were wholly inconsistent with his thesis and that his German authority was a book (by Bottler) on the dyeing of animal fibers by artificial dyes (including diazo compounds) and had nothing to do with natural animal pigments (apart from casual remarks on brown wool). I still have a reprint and have just read it.

I do not recall seeing any criticisms in print except Gortner's paper; "1911 Studies on Melanin IV. The origin of the pigment and color patterns in the elytra of the Colorado Potato Beetle (Leptinotarsa decemlineata). Amer. Nat. 45:743–755." There is no doubt, however, that Tower's work was discounted after 1911.

Sincerely yours,
Sewall Wright
Professor of Genetics

Tower's work may have contained some inconsistencies and some of his interpretations may have been erroneous, but there is no evidence that anyone else attempted to experiment with Leptinotarsa to confirm or disprove his findings. His work is cited in this review because, as Philiptchenko (1929, p. 110) expressed it, he was the first to investigate mutational changes in representatives of the animal kingdom. These animals happened to be species of Coleoptera.

References

Alexander, P., and Barton, D. H. R. 1943. The excretion of ethylquinone by the flour beetle. *Biochem. J.* **37**, 463–465.

Amer, N. M. 1963a. Modification of radiation effects with magnetic fields. *Univ. Calif. Lawrence Radiation Lab. Rept.* UCRL-11033.

Amer, N. M. 1963b. Modification of radiation injury with magnetic fields. *Radiation Res.* **19**, 179 (abstr.).

Amer, N. M., Slater, J. V., and Tobias, C. A. 1962. Analysis of the combined influence of X-irradiation and elevated temperatures on development. *Radiation Res.* **16**, 574 (abstr.).

Amos, T. G., and Scott, J. S. 1965. Pearl eye in *Carpophilus dimidiatus* F. (Col. Nitidulidae). *Am. Naturalist* **99**, 421–423.

Arendsen Hein, S. A. 1920. Studies on variation in the mealworm *Tenebrio molitor* L. I. Biological and genetical notes on *Tenebrio molitor* L. *J. Genet.* **10**, 227–263.

Arendsen Hein, S. A. 1924a. Studies on variation in the mealworm *Tenebrio molitor* L. II. Variations in tarsi and antennae. *J. Genet.* **14**, 1–38.

Arendsen Hein, S. A. 1924b. Selektionsversuche mit Prothorax- und Elytravariationen bei *Tenebrio molitor*. *Entomol. Mitt.* **153**, 243–275.

Balazuc, J. 1948. La tératologie des Coléoptères et expériences de transplantation sur *Tenebrio molitor* L. *Mem. Museum Natl. Hist. Nat.* (*Paris*) [N. S.] **25**, 1–293.

Bartlett, A. C. 1962. Section on new mutants. *Tribolium Inform. Bull.* **5**, 13.

Bartlett, A. C. 1964a. Two confirmed mutants in the boll weevil. *Ann. Entomol. Soc. Am.* **57**, 261–262.

Bartlett, A. C. 1964b. Section on new mutants. *Tribolium Inform. Bull.* **7**, 31.

Bartlett, A. C. 1965. Section on new mutants. *Tribolium Inform. Bull.* **8**, 40–41.

Bartlett, A. C., Bell, A. E., and Shideler, D. 1962a. Two loci controlling body color in flour beetles. *J. Heredity* **53**, 291–295.

Bartlett, A. C., Englert, D. C., and Blair, P. V. 1962b. Paddle versus wild type: a re-evaluation. *Tribolium Inform. Bull.* **5**, 23–24.

Bateson, W. 1894. "Materials for the Study of Variation," 598 pp. Macmillan, London.

Bateson, W. 1895. On the color variation of a beetle of the family Chrysomelidae, statistically examined. *Proc. Zool. Soc. London* **65**, 850–860.

Beck, J. S. 1962. Cell differentiation and radiopathology in the wing of *Tribolium confusum*. *Univ. Calif. Lawrence Radiation Lab. Rept.* **10211**, 122–159.

Beck, J. S. 1963. Effects of X-irradiation on cell differentiation and morphogenesis in a developing beetle wing. *Radiation Res.* **19**, 569–581.

Beck, J. S., and Slater, J. V. 1961. Effects of irradiation in *Tribolium confusum*. University of California Lawrence Radiation Laboratory, Report Biomedical Program Directors, U.S. Atomic Energy Commission Report 10.

Beck, J. S., and Manney, T. R. 1962. Neutron activation analysis for phosphorus in a study of development in a beetle wing. *Science* **137**, 865–866.

Bell, A. E. 1964. Section on new mutants. *Tribolium Inform. Bull.* **7**, 31–32.

Bell, A. E. 1965. Section on new mutants. *Tribolium Inform. Bell.* **8**, 41.

Bell, A. E., and Shideler, D. M. 1964. Tests of allelism of the scar (*sc*) and the engraved metasternum (*ems*) mutations in *Tribolium castaneum*. *Tribolium Inform. Bull.* **7**, 44–46.

Bell, A. E., Shideler, D. M., and Eddleman, H. L. 1964. Dominance and penetrance of the scar (*sc*) mutant in *Tribolium castaneum* as influenced by temperature. *Tribolium Inform. Bull.* **7**, 46–48.

Bender, H. A., and Doll, J. P. 1963. Bacteriologically sterile *Tribolium*. *Tribolium Inform. Bull.* **6**, 34.

Bray, D. F., Bell, A. E., and King, S. C. 1962. The importance of genotype by environment interaction with reference to control populations. *Genet. Res.* (*Cambridge*) **3**, 282–302.

Breitenbecher, J. K. 1921. The genetic evidence of a multiple allelomorph system in *Bruchus* and its relation to sex-limited inheritance. *Genetics* **6**, 65–90.

Breitenbecher, J. K. 1922. Somatic mutations and elytral mosaics of *Bruchus*. *Biol. Bull.* **43**, 10–22.

Breitenbecher, J. K. 1925. An apterous mutation in *Bruchus*. *Biol. Bull.* **48**, 166–170.

Breitenbecher, J. K. 1926. Variation and heredity in *Bruchus quadrimaculatus* Fabr. (Coleoptera). *Can. Entomologist* **58**, 131–133.

Bridges, C. R., and Dobzhansky, T. 1933. The mutant "proboscipedia" in *Drosophila melanogaster*—a case of hereditary homoösis. *Arch. Entwicklungsmech. Organ.* **127**, 575–590.

Brues, C. T., Melander, A. L., and Carpenter, F. M. 1954. Classification of insects. *Bull. Museum Comp. Zool. Harvard Coll.* **108**, 1–917.

Bywaters, J. H. 1960. Section on new mutants. *Tribolium Inform. Bull.* **3**, 24.

Castle, W. E. 1925. "Genetics and Eugenics," 3rd ed., 320pp. Harvard Univ. Press, Cambridge, Massachusetts.

Chapman, R. N. 1924. Nutritional studies on the confused flour beetle *Tribolium confusum* Duval. *J. Gen. Physiol.* **11**, 565–585.

Crenshaw, J. W., and Lerner, I. M. 1964. Productivity of inbred strains of *Tribolium confusum* and *Tribolium castaneum*. *Ecology* **45**, 697–705.

Crowson, R. A. 1960. The phylogeny of Coleoptera. *Ann. Rev. Entomol.* **5**, 111–134.

Daly, H. V., and Sokoloff, A. 1965. Labiopedia, an unusual mutant of *Tribolium confusum* Duval (Coleoptera: Tenebrionidae). *J. Morphol.* **117**, 251–270.

Dawson, P. S. 1961. A note on morphological variants in *Tribolium castaneum*. *Tribolium Inform. Bull.* **4**, 19.

Dawson, P. S. 1962a. Preliminary report on a possible phenodeviant in *T. castaneum*. *Tribolium Inform. Bull.* **5**, 24–25.

Dawson, P. S. 1962b. Section on new mutants. *Tribolium Inform. Bull.* **5**, 15.

Dawson, P. S. 1962c. Aberrant segregation ratios in *Tribolium castaneum*. *Records Genet. Soc. Am.* **31**, 81 and *Genetics* **47**, 950 (abstr.).

Dawson, P. S. 1963a. The time of action of lethality associated with the truncated elytra (*te*) gene in *T. castaneum*. *Tribolium Inform. Bull.* **6**, 34–35.

Dawson, P. S. 1963b. Somatic mutation in *T. castaneum* involving red. *Tribolium Inform. Bull.* **6**, 37–38.

Dawson, P. S. 1964a. Section on new mutants. *Tribolium Inform. Bull.* **7**, 31–43.

Dawson, P. S. 1964b. The genetics of developmental rate and its relationship to competitive ability in *Tribolium*. Ph.D. thesis, University of California, Berkeley, California. 167pp.

Dawson, P. S. 1964c. "Pokey": a sex-linked recessive semi-lethal gene in *Tribolium castaneum* which greatly prolongs the larval stage of development. *Z. Vererbungslehre* **95**, 215–221.

Dawson, P. S. 1965. "Serrate": a sex-linked recessive gene in the flour beetle, *Tribolium castaneum*. *Can. J. Genet. Cytol.* **7**, 559–562.

Dawson, P. S., and Ho, F. K. 1962. Section on new mutants. *Tribolium Inform. Bull.* **5**, 15.

Dawson, P. S., and Lerner, I. M. 1062. Genetic variation and indeterminism in interspecific competition. *Am. Naturalist* **96**, 379–380.

Dawson, P. S., and Sokoloff, A. 1964. A multiple allelic series in *Tribolium castaneum*. *Am. Naturalist* **98**, 455–457.

Dewees, A. A. 1963. Frequency and relative fitness of some mutants involving the eye and body color in a natural population of *Tribolium castaneum*. M. A. thesis, Southern Illinois University, Carbondale, Illinois. 31pp.

Dewees, A. A. 1965. Linkage information on peach and ruby. *Tribolium Inform. Bull.* **8**, 72.

Dobzhansky, T. 1924. Die geographische und individuelle Variabilität von *Harmonia axyridis* Pall. in ihren Wechselbeziehungen. *Biol. Zentr.* **44**, 401–421.

Dobzhansky, T. 1933. Geographical variation in lady-beetles. *Am. Naturalist* **67**, 97–126.

Dyte, C. E. 1963. Section on stocks. *Tribolium Inform. Bull.* **6**, 16–20.

Dyte, C. E. 1964. Section on stock lists. *Tribolium Inform. Bull.* **7**, 24.

Dyte, C. E., and Blackman, D. G. 1961. A pearl-eyed mutation in *Gnathocerus cornutus* (Fab.) (Coleoptera: Tenebrionidae), with notes on similar variants in other beetles. *Proc. Roy. Entomol. Soc. London* A **36**, 168–172.

Dyte, C. E., and Blackman, D. G. 1962a. Ebony-2: a new mutant in *T. confusum*. *Tribolium Inform. Bull.* **8**, 41–42.

Dyte, C. E., and Blackman, D. G. 1962b. A new black mutation linked to "pearl" in *Tribolium confusum* Duval (Coleoptera: Tenebrionidae). *Am. Naturalist* **96**, 376–378.

Dyte, C. E., Shaw, D. D., and Blackman, D. G. 1965. Section on new mutants. *Tribolium Inform. Bull.* **8**, 41–42.

Eddleman, H. L. 1961. Section on new mutants. *Tribolium Inform. Bull.* **4**, 14.

Eddleman, H. L. 1962. Section on new mutants. *Tribolium Inform. Bull.* **5**, 13–14.

Eddleman, H. L. 1964. Section on new mutants. *Tribolium Inform. Bull.* **7**, 31.

Eddleman, H. L., and Bell, A. E. 1963. Four new eye color mutants in *Tribolium castaneum*. *Genetics* **48**, 888 (abstr.).

El Kifl, A. H. 1953. Morphology of the adult *Tribolium confusum* Duv. and its differentiation from *Tribolium* (*Stene*) *castaneum* Herbst (Coleoptera: Tenebrionidae). *Bull. Soc. Fouad Ier Entomol.* **37**, 173–249.

Engelhardt, M., Rapoport, H., and Sokoloff, A. 1965. Comparison of the content of the odoriferous gland reservoirs in normal and mutant *Tribolium confusum*. *Science* **150**, 632–633.

Englert, D. C. 1963. Section on new mutants. *Tribolium Inform. Bull.* **6**, 24.

Englert, D. C. 1964. The genetics of growth in the flour beetle *Tribolium castaneum*. Ph.D. thesis, Purdue University, Lafayette, Indiana. 130pp.

Englert, D. C., and Bell, A. E. 1963a. "Antennapedia" and "squint," recessive marker genes for linkage group VIII in *Tribolium castaneum. Can. J. Genet. Cytol.* **5**, 467–471.

Englert, D. C., and Bell, A. E. 1963b. "Antennapedia": an unusual antennal mutation in *Tribolium castaneum. Ann. Entomol. Soc. Am.* **56**, 123–124.

Englert, D. C., Shideler, D., and Bell, A. E. 1963. "Antennapedia" and "squint," recessive marker genes for linkage group VIII in *Tribolium castaneum. Genetics* **48**, 888–889.

Ferwerda, F. P. 1928. Genetische Studien am Mehlkäffer. *Genetica* **11**, 1–111.

Gadeau de Kerville, H. 923. Description et figuration d'anomalies coléoptèrologiques. *Bull. Soc. Entomol. France* **18**, 229–233.

Goldschmidt, R. 1938. "Physiological Genetics," 375pp. McGraw-Hill, New York.

Goldschmidt, R. 1940. "The Material Basis of Evolution," 436pp. Yale Univ. Press, New Haven, Connecticut.

Good, N. E. 1936. The flour beetles of the genus *Tribolium. U. S. Dept. Agr. Tech. Bull.* **498**, 1–58.

Graham, W. M. 1957. Pearl eye in the confused flour beetle *Tribolium confusum* Duval (Tenebrionidae). *Entomologist's Monthly Mag.* **93**, 73–75.

Gregory, W. K. 1946. The roles of motile larvae and fixed adults in the origin of vertebrates. *Quart. Rev. Biol.* **21**, 348–364.

Hackman, R. H., Pryor, M. G. M., and Todd, A. R. 1948. The occurrence of phenolic substances in arthropods. *Biochem. J.* **43**, 474–477.

Haldane, J. B. S. 1922. Sex ratio and unisexual sterility in hybrid animals. *J. Genet.* **12**, 101–109.

Hinton, H. E. 1942. Secondary sexual characters of *Tribolium. Nature* **149**, 500.

Hinton, H. E. 1948. A synopsis of the genus *Tribolium* Macleay with some remarks on the evolution of its species groups. *Bull. Entomol. Res.* **39**, 13–55.

Ho, F. K. 1961. Optic organs of *Tribolium confusum* and *T. castaneum* and their usefulness in age determination (Coleoptera, Tenebrionidae). *Ann. Entomol. Soc. Am.* **54**, 921–925.

Ho, F. K. 1962a. Section on new mutants. *Tribolium Inform. Bull.* **5**, 17–18.

Ho, F. K. 1962b. Occurrence of "pearl" in "natural" and laboratory populations of *Gnathocerus cornutus. Tribolium Inform. Bull.* **5**, 28–29.

Ho, F. K. 1963. Preliminary studies of cannibalism in *Tribolium. Tribolium Inform. Bull.* **6**, 39–40.

Ho, F. K., and Dawson, P. S. 1962. Section on new mutants. *Tribolium Inform. Bull.* **5**, 15.

Ho, F. K., and Sokoloff, A. 1962. Occurrence of mutations in "natural" populations of *Tribolium. Tribolium Inform. Bull.* **5**, 29–30.

Hosino, Y. 1940. Genetical studies on the pattern types of the lady-bird beetle, *Harmonia axyridis* Pallas. *J. Genet.* **40**, 215–228.

Howe, R. W. 1956. The effect of temperature and humidity on the rate of development and mortality of *Tribolium castaneum* (Herbst) (Coleoptera, Tenebrionidae). *Ann Appl. Biol.* **44**, 356–368.

Howe, R. W. 1960. The effect of temperature and humidity on the rate of development of *Tribolium confusum* Duval (Coleoptera, Tenebrionidae). *Ann. Appl. Biol.* **48**, 363–376.

Howe, R. W. 1962a. The effect of temperature and relative humidity on the rate of development of *Tribolium madens* (Charp.) (Coleoptera, Tenebrionidae). *Ann. Appl. Biol.* **50**, 649–660.

Howe, R. W. 1962b. Section on stock lists. *Tribolium Inform. Bull.* **5,** 11.

Imms, A. D. 1940. On the growth processes in the antennae of insects. *Quart. J. Microscop. Sci.* **81,** 585–593.

Jayne, H. F. 1880. Descriptions of some monstrosities observed in North American Coleoptera. *Trans. Am. Entomol. Soc.* **8,** 155–162.

Kollros, C. L. 1944. A study of the gene, *pearl,* in populations of *Tribolium castaneum* Herbst. Unpublished Ph.D. dissertation. The University of Chicago Libraries, Chicago, Illinois. 61pp.

Komai, T. 1956. Genetics of ladybeetles. *Advan. Genet.* **8,** 155–188.

Komai, T., Chino, M., and Hosino, Y. 1950. Contributions to the evolutionary genetics of the lady-beetle, *Harmonia.* I. Geographic and temporal variations in the relative frequencies of the elytral pattern types and in the frequency of the elytral ridge. *Genetics* **35,** 589–601.

Krause, E. 1963a. Rate of development of the *Sa* mutant. *Tribolium Inform. Bull.* **6,** 44–45.

Krause, E. 1963b. Effect of temperature on penetrance of the *ti* mutant. *Tribolium Inform. Bull.* **6,** 44.

Krause, E., Shideler, D., and Bell, A. E. 1962. An autosomal dominant mutant in *Tribolium castaneum* with recessive lethal effects. *Am. Naturalist* **96,** 186–188.

Lasley, E. L. 1960a. Evidence for equality of recombination between split and jet in both sexes of *Tribolium castaneum. Tribolium Inform. Bull.* **3,** 14–15.

Lasley, E. L. 1960b. Section on new mutants. *Tribolium Inform. Bull.* **3,** 24.

Lasley, E. L. 1960c. The incompletely recessive effect of the sex-linked gene, pygmy, on pupa weight in *Tribolium castaneum. Tribolium Inform. Bull.* **3,** 21–22.

Lasley, E. L., and Sokoloff, A. 1960. Section on new mutants. *Tribolium Inform. Bull.* **3,** 22.

Lasley, E. L., and Sokoloff, A. 1961. Section on new mutants. *Tribolium Inform. Bull.* **4,** 16.

Leclercq, J. 1963. Artificial selection for weight and its consequences in *Tenebrio molitor* L. *Nature* **198,** 106–107.

Lefkovitch, L. P. 1963. Differing status of color forms in *Cryptolestes* Gangl. (Cucujidae). *Tribolium Inform. Bull.* **7,** 45–46.

Lerner, I. M., and Dempster, E. R. 1962. Indeterminism in interspecific competition. *Proc. Natl. Acad. Sci. U.S.* **48,** 821–826.

Lerner, I. M., and Ho, F. K. 1960. Black mutant of *T. confusum. Tribolium Inform. Bull.* **3,** 14.

Lerner, I. M., and Ho, F. K. 1961. Genotype and competitive ability of *Tribolium* species. *Am. Naturalist* **95,** 329–343.

Loconti, J. D., and Roth, L. M. 1953. Composition of the odorous secretion of *Tribolium castaneum. Ann. Entomol. Soc. Am.* **46,** 281–289.

Lus, J. 1928. On the inheritance of color and pattern in lady beetles *Adalia bipunctata* L. and *Adalia decempunctata* L. *Bull. Bur. Genet., Leningrad* **6,** 89–163.

McCracken, I. 1905. A study of the inheritance of dichromatism in *Lina lapponica. J. Exptl. Zool.* **2,** 117–137.

McCracken, I. 1906. Inheritance of dichromatism in *Lina* and *Gastroidea. J. Exptl. Zool.* **3,** 321–336.

McCracken, I. 1907. Occurrence of a sport in *Melasoma* (*Lina*) *scripta* and its behavior in heredity. *J. Exptl. Zool.* **4,** 221–239.

McDonald, D. J. 1959a. Two new deleterious mutations of *Tribolium confusum*. *J. Heredity,* **50**, 85–88.

McDonald, D. J. 1959b. A study of mutations affecting the viability of *Tribolium confusum*. *Proc. 10th Intern. Congr. Genet., Montreal, 1958*. Vol. 2, pp. 183–184. Univ. of Toronto Press, Toronto, Ontario, Canada.

McDonald, D. J., and Peer, N. J. 1960. Natural selection in experimental populations of *Tribolium* I. Preliminary experiments with population cages. *Genetics* **45**, 1317–1333.

McDonald, D. J., and Peer, N. J. 1961a. Natural selection in experimental populations of *Tribolium*. II. Factors affecting the fitness of the mutant "split" of *Tribolium confusum*. *Heredity* **16**, 317–330.

McDonald, D. J., and Peer, N. J. 1961b. Section on new mutants. *Tribolium Inform. Bull.* **4**, 14.

McDonald, D. J., Spencer, C., and Wagner, M. 1963. "Blistering" in *Tribolium confusum*. *J. Heredity* **54**, 183–186.

Miller, L. W. 1944. Investigations of the flour beetles of the genus *Tribolium*. A color strain of *T. castaneum* (Hbst.). *J. Dept. Agr. Victoria* **42**, 469–471.

Mocquerys, S. 1880. Recueil de Coléoptères anormaux par feu M. S. Mocquerys avec introduction par M. J. Bourgeois. 142pp. L. Deshays, Rouen, France.

Neboiss, A. 1962. Notes on the distribution and description of new species (Orders: Odonata, Plecoptera, Orthoptera, Trichoptera and Coleoptera). *Mem. Melbourne Nat. Museum* **25**, 243–258.

Newman, H. H. 1932. "Evolution, Genetics and Eugenics." Univ. of Chicago Press, Chicago, Illinois.

Neyman, J., Park, T., and Scott, E. L. 1955. Struggle for existence. The Tribolium model. Biological and statistical aspects. *Proc. 3rd Berkeley Symp. Math. Statist. Probability, 1955* pp. 41–79.

Park, T. 1934a. Observations on the general biology of the flour beetle *Tribolium confusum*. *Quart. Rev. Biol.* **9**, 36–54.

Park, T. 1934b. Studies in population physiology. III. The effect of conditioned flour upon the productivity and population decline of *Tribolium confusum*. *J. Exptl. Zool.* **68**, 167–182.

Park, T. 1935. Studies on population physiology. IV. Some physiological effects of conditioned flour upon *Tribolium confusum* Duv. and its population. *Physiol. Zool.* **8**, 91–115.

Park, T. 1936. Studies in population physiology. VI. The effect of differentially conditioned flour upon the fecundity and fertility of *Tribolium confusum* Duval. *J. Exptl. Zool.* **73**, 393–404.

Park, T. 1937. The inheritance of the mutation "pearl" in the flour beetle *Tribolium castaneum* Herbst. *Am. Naturalist* **71**, 143–157.

Park, T. 1948. Experimental studies of inter-species competition. I. Competition between populations of the flour beetles *Tribolium confusum* Duval, and *Tribolium castaneum* Herbst. *Ecol. Monographs* **18**, 265–308.

Park, T. 1954. Experimental studies of inter-species competition. II. Temperature, humidity, and competition in two species of *Tribolium*. *Physiol. Zool.* **27**, 177–238.

Park, T. 1957. Experimental studies of inter-species competition. III. Relation of initial species proportion to competitive outcome in populations of *Tribolium*. *Physiol. Zool.* **30**, 22–40.

Park, T., and Frank, M. B. 1951. "Paddle": a sex-linked recessive gene of *Tribolium castaneum* Herbst. *Am. Naturalist* **85**, 313–318.

Park, T., and Lloyd, M. 1955. Natural selection and the outcome of competition. *Am. Naturalist* **89**, 235–240.

Park, T., and Woollcott, N. 1937. Studies in population physiology. VII. The relation of environmental conditioning to the decline of *Tribolium confusum* populations. *Physiol. Zool.* **10**, 197–211.

Park, T., Ginsburg, B., and Horowitz, S. 1945. Ebony: a gene affecting the body color and fecundity of *Tribolium confusum* Duval. *Physiol. Zool.* **18**, 35–52.

Park, T., DeBruyn, P. P. H., and Bond, J. A. 1958. The relation of X-irradiation to the fecundity and fertility of two species of flour beetles. *Physiol. Zool.* **31**, 151–170.

Park, T., Mertz, D. B., and Petrusewicz, K. 1961. Genetic strains of *Tribolium*, their primary characteristics. *Physiol. Zool.* **34**, 62–80.

Park, T., Leslie, P. H., and Mertz, D. B. 1964. Genetic strains and competition in populations of *Tribolium*. *Physiol. Zool.* **37**, 97–162.

Philip, V. 1940. A genetical analysis of three small populations of *Dermestes vulpinus* F. (Coleoptera). *Proc. Indian Acad. Sci.* **12**, 133–171.

Philiptchenko, Yur. A. 1929. "Genetics," Government Edition. 397pp. Moscow.

Phillips, A. L., and McDonald, D. J. 1958. A comparison of productivity in a mutant and wild strain of *Tribolium castaneum* Herbst. *Am. Naturalist* **92**, 376–378.

Pohley, H. J. 1959. Experimentelle Beiträge zur Lenkung der Organentwicklung, des Häutungsrhythmus und der Metamorphose bei der Schabe *Periplaneta americana*. *Arch. Entwicklungsmech. Organ.* **151**, 323–380.

Prus, T. 1961. The effect of homotypic and heterotypic conditioning of medium upon the net fecundity of *Tribolium castaneum* Herbst and *T. confusum* Duval. *Ekologia Polska Ser. A.* **9**, 245–247.

Przibram, H. 1919. Fussglieder an Käferfühlern (Zugleich Homoesis bei Arthropoden, 5-Mitteilung). *Arch. Entwicklungsmech. Organ.* **45**, 52–68.

Reynolds, S. 1964. Section on new mutants. *Tribolium Inform. Bull.* **7**, 32.

Reynolds, E. M., and Sylvester, N. K. 1961. Pearl mutant in *Trogoderma granarium* Everts. (Dermestidae). *Tribolium Inform. Bull.* **4**, 25.

Ridgway, R. 1912. "Color Standards and Color Nomenclature," 43pp. A. Hoen, Washington, D. C

Roth, L. M. 1943. Studies on the gaseous secretion of *Tribolium confusum* Duval. II. The odoriferous glands of *Tribolium confusum*. *Ann. Entomol. Soc. Am.* **36**, 397–424.

Roth, L. M., and Eisner, T. 1962. Chemical defenses of arthropods. *Ann. Rev. Entomol.* **7**, 107–136.

Roth, L. M., and Howland, R. B. 1941. Studies on the gaseous secretion of *T. confusum* Duval. I. Abnormalities produced in *T. confusum* Duv. by exposure to a secretion given off by adults. *Ann. Entomol. Soc. Am.* **34**, 151–172.

Roth, L. M., and Stay, B. 1958. The occurrence of *para*-quinones in some arthropods, with emphasis on the quinone-secreting tracheal glands of *Diploptera punctata* (Blattaria). *J. Insect Physiol.* **1**, 305–318.

Schlager, G. 1963. The ecological genetics of the mutant, sooty, in populations of *Tribolium castaneum*. *Evolution* **17**, 254–273.

Schuurman, J. J. 1937. Contributions to the genetics of *Tenebrio molitor* L. *Genetica* **19**, 273–355.

Shaw, D. D. 1965. Pleiotropic effects of mutants affecting eye-colour in seven species of Coleoptera. *Tribolium Inform. Bull.* **8**, 128.

Shideler, D. M. 1962. Section on new mutants. *Tribolium Inform. Bull.* **5**, 13.

Shull, A. F. 1943. Inheritance in lady beetles. I. The spotless and spotted elytra in *Hippodamia sinuata. J. Heredity* **34**, 329–337.

Shull, A. F. 1944. Inheritance in lady beetles. II. The spotless pattern and its modifiers in *Hippodamia convergens* and their frequency in several populations. *J. Heredity* **35**, 329–339.

Shull, A. F. 1945. Inheritance in lady beetles. III. Crosses between variants of *Hippodamia quinquesignata* and between this species and *H. convergens. J. Heredity* **36**, 149–160.

Shull, A. F. 1946. The form of the chitinous male genitalia in crosses of the species *Hippodamia quinquesignata* and *H. convergens. Genetics* **31**, 291–303.

Slater, J. V., Rescigno, A., Amer, N. M., and Tobias, C. A. 1963. Temperature dependence of wing abnormality in *Tribolium confusum. Science* **140**, 408–409.

Slater, J. V., Lyman, J., Tobias, C. A., Amer, N. M., Beck, J. S., Beck, M., and Slater, A. J. 1964. Heavy ion localization of sensitive embryonic sites in *Tribolium. Radiation Res.* **21**, 541–549.

Slater, J. V., Tobias, C. A., Beck, J. S., Lyman, J. T., Martin, M. E., and Luce, J. R. 1961. Comparative influence of accelerated heavy nuclei on anomalous development in *Tribolium. Radiation Res.* **14**, 503–504.

Smith, S. G. 1949. Evolutionary changes in the sex chromosomes of Coleoptera. I. Wood borers of the genus *Agrilus. Evolution* **3**, 344–357.

Smith, S. G. 1950. The cytotaxonomy of Coleoptera. *Can. Entomologist.* **82**, 58–68.

Smith, S. G. 1951. Evolutionary changes in the sex chromosomes of Coleoptera. *Genetica* **25**, 522–524.

Smith, S. G. 1952a. The evolution of heterochromatin in the genus *Tribolium* (Tenebrionidae: Coleoptera). *Chromosoma* **4**, 585–610.

Smith, S. G. 1952b. The cytology of some tenebrionoid beetles (Coleoptera). *J. Morphol.* **91**, 325–364.

Smith, S. G. 1953. Chromosome numbers of Coleoptera. *Heredity* **7**, 31–48.

Smith, S. G. 1958. Animal cytology and Cytotaxonomy. *Proc. Genet. Soc. Can.* **3**, 57–64.

Smith, S. G. 1959. The cytogenetic basis of speciation in Coleoptera. *Proc. Intern. Congr. Genet. 10th, Montreal. 1958* Vol. I, pp. 444–450. Univ. of Toronto Press, Toronto, Ontario, Canada.

Smith, S. G. 1960a. Cytogenetics of insects. *Ann. Rev. Entomol.* **5**, 69–84.

Smith, S. G. 1960b. Chromosome numbers of Coleoptera II. *Can. J. Genet. Cytol.* **2**, 66–88.

Snow, R. 1962. The chromosome number of *Gnathocerus cornutus. Tribolium Inform. Bull.* **5**, 39.

Sokal, R. R., and Huber, I. 1963. Competition among genotypes in *Tribolium castaneum* at varying densities and gene frequencies (the sooty locus). *Am. Naturalist* **97**, 169–184.

Sokal, R. R., and Karten, I. 1964. Competition among genotypes in *Tribolium castaneum* at varying densities and gene frequencies (the black locus). *Genetics* **49**, 195–211.

Sokal, R. R., and Sonleitner, F. J. 1965. Components of selection in *Tribolium* (Coleoptera) and houseflies. *Proc. 12th Intern. Congr. Entomol., London, 1964.* Vol. I, pp. 274–275. Roy. Entomol. Soc., London.

Sokoloff, A. 1959. The nature of the "pearl" mutation in *Tribolium castaneum* and *Latheticus oryzae* (Tenebrionidae). *Anat. Record* **134**, 641–642.

Sokoloff, A. 1960a. Linkage studies in *Tribolium castaneum* Herbst. III. A preliminary report on "truncated elytra," a sex-linked recessive gene with lethal effects. *Can. J. Genet. Cytol.* **2**, 379–388.

Sokoloff, A. 1960b. Aberrations in *Tribolium* and *Latheticus oryzae*. *Tribolium Inform. Bull.* **3**, 30–34.

Sokoloff, A. 1960c. Section on new mutants. *Tribolium Inform. Bull.* **3**, 22–25.

Sokoloff, A. 1960d. Linkage studies in *Tribolium castaneum* Herbst. I. Recombination between paddle and miniature appendaged. *Can. J. Genet. Cytol.* **2**, 28–33.

Sokoloff, A. 1961a. Irradiation experiments with *Tribolium*. *Tribolium Inform. Bull.* **4**, 28–33.

Sokoloff, A. 1961b. Section on new mutants. *Tribolium Inform. Bull.* **4**, 15–18.

Sokoloff, A. 1962a. Linkage studies in *Tribolium castaneum* Herbst. V. The genetics of Bar eye, microcephalic, and Microphthalmic and their relationships to black, jet, pearl and sooty. *Can. J. Genet. Cytol.* **4**, 409–425.

Sokoloff, A. 1962b. Chromosome maps of *Tribolium castaneum* with suggestions for designation of chromosome numbers in other tenebrionids. *Tribolium Inform. Bull.* **5**, 42–44a.

Sokoloff, A. 1962c. A simple technique for ridding *Tribolium* cultures of parasites. *Tribolium Inform. Bull.* **5**, 48.

Sokoloff, A. 1962d. Section on new mutants. *Tribolium Inform. Bull.* **5**, 14–19.

Sokoloff, A. 1962e. Linkage studies in *Tribolium castaneum* Herbst. IV. Further data on the position of "truncated elytra." *Can. J. Genet. Cytol.* **4**, 133–140.

Sokoloff, A. 1963a. Further linkage results for *Tribolium castaneum*. *Genetics* **48**, 910–911 (abstr.).

Sokoloff, A. 1963b. Unusual mutants in *Tribolium castaneum* and *T. confusum*. *Genetics* **48**, 910 (abstr.).

Sokoloff, A. 1963c. Studies on factors affecting crossing over in *Tribolium castaneum*. *Tribolium Inform. Bull.* **6**, 57–60.

Sokoloff, A. 1963d. A somatic mutation involving squint (*sq*). *Tribolium Inform. Bull.* **6**, 56–57.

Sokoloff, A. 1963e. Section on new mutants. *Tribolium Inform. Bull.* **6**, 23–32.

Sokoloff, A. 1963f. Linkage studies in *Tribolium castaneum* Herbst. VI. "Divergent elytra," an incompletely recessive sex-linked gene. *Can. J. Genet. Cytol.* **5**, 12–17.

Sokoloff, A. 1964a. A dominant synthetic lethal in *Tribolium castaneum* Herbst. *Am. Naturalist* **98**, 127–128.

Sokoloff, A. 1964b. Sex and crossing over in *Tribolium castaneum*. *Genetics* **50**, 491–496.

Sokoloff, A. 1964c. Two abnormalities associated with microcephalic. *Tribolium Inform. Bull.* **7**, 78.

Sokoloff, A. 1964d. Three unusual aberrations in *Tribolium castaneum*. *Tribolium Inform. Bull.* **7**, 77–78.

Sokoloff, A. 1964e. Revised maps of *Tribolium castaneum*, *T. confusum* and *Latheticus oryzae*. *Tribolium Inform. Bull.* **7**, 72–76.

Sokoloff, A. 1964f. Linkage studies in *Tribolium confusum*. I. Preliminary studies with autosomal genes. *Can. J. Genet. Cytol.* **6**, 259–270.

Sokoloff, A. 1964g. Linkage studies in *Latheticus oryzae* Waterh. II. Linkage of "brown body" and "creased abdominal sternites" with pearl. *Can. J. Genet. Cytol.* **6**, 271–276.

Sokoloff, A. 1964h. Section on new mutants. *Tribolium Inform. Bull.* **7**, 31–43.

Sokoloff, A. 1964i. Linkage studies in *Tribolium castaneum* Herbst. X. Light ocular diaphram, a gene linked with black. *Can. J. Genet. Cytol.* **6**, 147–151.

Sokoloff, A. 1965a. An unusual modifier-suppressor system in *Tribolium castaneum. Am. Naturalist* **99**, 143–151.

Sokoloff, A. 1965b. Revised linkage maps in *Tribolium castaneum* and *T. confusum. Tribolium Inform. Bull.* **8**, 141–144.

Sokoloff, A. 1965c. Section on new mutants. *Tribolium Inform Bull.* **8**, 43–64.

Sokoloff, A. 1965d. Studies of natural selection in *Tribolium castaneum* Herbst (Coleoptera) in two different media. *Proc. 12th Intern. Congr. Entomol., London, 1964.* Vol. I, pp. 273–274. Entomol. Soc., London.

Sokoloff, A., and Dawson, P. S. 1963a. Linkage studies in *Tribolium castaneum* Herbst. IX. The map position of antennapedia, squint, short elytra and elbowed antenna. *Can. J. Genet. Cytol.* **5**, 450–458.

Sokoloff, A., and Dawson, P. S. 1963b. Linkage studies in *Tribolium castaneum* Herbst VII. The map position of four sex-linked lethals. *Can. J. Genet. Cytol.* **5**, 138–145.

Sokoloff, A., and Ho, F. K. 1963. Section on new mutants. *Tribolium Inform. Bull.* **6**, 23–32.

Sokoloff, A., and Hoy, M. A. 1965. Possible genetic basis for prothetely in *Tribolium castaneum* and *Latheticus oryzae. Tribolium Inform. Bull.* **8**, 150–151.

Sokoloff, A., and Lasley, E. L. 1961. Section on new mutants. *Tribolium Inform. Bull.* **4**, 15–18.

Sokoloff, A., and Shrode, R. R. 1960. Linkage studies in *Latheticus oryzae* Waterh. I. Recombination between "red" and "truncated elytra." *Can. J. Genet. Cytol.* **2**, 418–428.

Sokoloff, A., and Shrode, R. R. 1962. Survival of *Tribolium castaneum* Herbst after rocket flight into the ionosphere. *Aerospace Med.* **33**, 1304–1317.

Sokoloff, A., Lasley, E. L., and Shrode, R. R. 1960a. Linkage studies in *Tribolium castaneum* Herbst. II. "Pygmy," "red," and "spotted" and their relationships to the *ma* and *pd* genes. *Can. J. Genet. Cytol.* **2**, 142–149.

Sokoloff, A., Slatis, H. M., and Stanley, J. 1960b. The black mutation in *Tribolium castaneum. J. Heredity.* **52**, 131–135.

Sokoloff, A., Dawson, P. S., and Englert, D. C. 1963. Linkage studies in *Tribolium castaneum* Herbst. VIII. Short antenna, a dominant marker for the seventh linkage group. *Can. J. Genet. Cytol.* **5**, 299–306.

Sokoloff, A., Shrode, R. R., and Bywaters, J. H. 1965. Productivity in *Tribolium castaneum. Physiol. Zool.* **38**, 165–173.

Sonleitner, F. J. 1961. Factors affecting egg cannibalism and fecundity in populations of adult *Tribolium castaneum* Herbst. *Physiol. Zool.* **34**, 233–255.

Stanley, J. 1961a. Two techniques of use in the control of *Triboliocystis garnhami. Can. J. Zool.* **39**, 121–122.

Stanley, J. 1961b. Sterile crosses between mutations of *Tribolium confusum* Duv. *Nature* **191**, 934.

Stanley, J. 1964a. A mathematical theory of the growth of populations of the flour beetle *Tribolium confusum* Duval. VIII. A further study of the "re-tunneling" problem. *Can. J. Zool.* **42**, 201–227.

Stanley, J. 1964b. Washing the eggs of *Tribolium* for gregarine control. *Can. J. Zool.* 42, 920.

Stanley, J., and Slatis, H. M. 1955. Studies from the autotrephon. IV. A black mutation of *T. confusum* compared with the normal reddish-brown strain. *Ecology* 36, 473–485.

Stanley, M. S. M. 1965. Section on new mutants. *Tribolium Inform. Bull.* 8, 64.

Tan, C. C., and Li, J. C. 1932–1933. *Peking Nat. Hist. Bull.* 7, 175–193.

Tan, C. C., and Li, J. C. 1934. Inheritance of the elytral color patterns of the lady-bird beetle *Harmonia axyridis* Pallas. *Am. Naturalist* 68, 252–265.

Tanaka, Y. 1953. Genetics of the silkworm *Bombyx nori*. *Advan. Genet.* 5, 239–317.

Timberlake, P. H. 1922. Observation on the phenomena of heredity in the ladybeetle *Coelophora inaequalis* (Fabricius). *Proc. Hawaii Entomol. Soc.* 5, 121–133.

Timofeeff-Ressovsky, H. 1935. Divergens, eine Mutation von *Epilachna chrysomelina*. F. Z. *Induktive Abstammungs- Vererbungslehre*, 68, 443–453.

Timofeeff-Ressovsky, N. W. 1940. Zur Analyse des Polymorphismus bei *Adalia bipunctata*. *Biol. Zentr.* 60, 130–137.

Tower, W. L. 1903. Colors and color patterns in Coleoptera. *Decennial Publ. Univ. Chicago. First Series* pp. 31–70.

Tower, W. L. 1906. An investigation of evolution in chrysomelid beetles of the genus *Leptinotarsa*. *Carnegie Inst. Wash. Publ.* 48, 1–320.

Tower, W. L. 1910a. The determination of dominance and the modification of behavior in alternative (Mendelian) inheritance, by conditions surrounding or incident upon the germ cells at fertilization. *Biol. Bull.* 20, 67–69.

Tower, W. L. 1910b. The determination of dominance and the modification of behavior in alternative (Mendelian) inheritance, by conditions surrounding or incident upon the germ cells at fertilization. *Biol. Bull.* 18, 285–352.

Van Wyk, J. H., Hodson, A. C., and Christensen, C. M. 1959. Microflora associated with the confused flour beetle, *Tribolium confusum*. *Ann. Entomol. Soc. Am.* 52, 452–453.

Villee, C. A. 1942. The phenomenon of homoeosis. *Am. Naturalist* 76, 494–506.

Villee, C. A. 1945. On homoeosis in Drosophila. *J. Elisha Mitchell Sci. Soc.* 61, 291–300.

von Heyden, L. 1881. Monstrose Käfer aus meiner und der Sammlung des H. Prof. Doebner in Aschaffenburg. *Deut. Entomol. Z.* 25, 105–110.

von Lengerken, H. 1933. Uber bilaterale und Pseudo-Fühler heteromorphose unter natürlicher Bedingungen. *Biol. Zentr.* 53, 1–10.

Waddington, C. H. 1940. "Organizers and Genes," 160 pp. Cambridge Univ. Press, London and New York.

Waddington, C. H., and Perry, M. M. 1963. An electron microscope analysis of eye development in *Tribolium castaneum*. *Tribolium Inform. Bull.* 6, 67–68.

Whittinghill, M. 1947. Spermatogonial crossing over between the third chromosomes in the presence of the Curly inversions in Drosophila melanogaster, *Genetics* 32, 608–614.

Wolsky, A., and Zamora, R. 1960. The structure of the "pearl" eye of *Latheticus oryzae*. *Am. Naturalist* 94, 309–312.

Yamada, Y. 1962. Section on new mutants. *Tribolium Inform. Bull.* 5, 13.

Zulueta, A. de. 1925. La herencia ligada al sexo en el coleóptero *Phytodecta variabilis* Ol. *Eos* (*Madrid*) 1, 203–229.

Zulueta, A. de 1932. Heredity in the X and Y chromosomes of *Phytodecta*. *Proc. 6th Intern. Congr. Genet., Ithaca, 1932.* Vol. 1, p. 389. Cornell Univ. Press, Ithaca, New York.

Author Index

Numbers in italics refer to pages on which the complete references are listed.

A

Alexander, P., 71, *185*
Amer, N. M., 63, *185*, *192*
Amos, T. G., 114, 161, *185*
Arendsen Hein, S. A., 116, 117, *185*

B

Balazuc, J., 76, 162, *185*
Bartlett, A. C., 25, 48, 132, 161, 162, *185*
Barton, D. H. R., 71, *185*
Bateson, W., 127, 151, *185*
Beck, J. S., 63, *185*, *186*, *192*
Beck, M., 63, *192*
Bell, A. E., 15, 25, 36, 39, 41, 42, 43, 49, 52, 71, 167, *186*, *187*, *188*, *189*
Bender, H. A., 3, *186*
Blackman, D. G., 93, 112, 113, 114, 121, 140, 154, 161, *187*
Blair, P. V., 11, *185*
Bond, J. A., 2, *191*
Bray, D. F., 167, *186*
Breitenbecher, J. K., 130, 131, 161, *186*
Bridges, C. R., 81, 152, *186*
Brues, C. T., 111, *186*
Bywaters, J. H., 2, 13, 43, 171, *186*, *194*

C

Carpenter, F. M., 111, *186*
Castle, W. E., 181, *186*
Chapman, R. N., 1, *186*
Chino, M., 111, *189*
Christensen, C. M., 106, *195*
Crenshaw, J. W., 171, *186*
Crowson, R. A., 163, *186*

D

Daly, H. V., 81, 151, *186*
Dawson, P. S., 15, 17, 18, 19, 20, 21, 25, 40, 41, 42, 43, 45, 50, 58, 61, 67, 69, 80, 101, 135, 139, 150, 156, 162, 166, 170, 171, *186*, *187*, *188*, *194*
DeBruyn, P. P. H., 2, *191*
Dempster, E. R., 165, 166, *189*
Dewees, A. A., 24, 25, 48, 49, *187*
de Zulueta, A., 128, *195*
Dobzhansky, T., 81, 111, 115, 152, *186*, *187*
Doll, J. P., 3, *186*
Dyte, C. E., 24, 93, 111, 112, 113, 114, 119, 121, 140, 154, 161, *187*

E

Eddleman, H. L., 15, 24, 35, 36, 39, 49, 52, 57, 58, 140, *186*, *187*
Eisner, T., 106, *191*
El Kifl, A. H., 31, 61, *187*
Engelhardt, M., 106, *187*
Englert, D. C., 11, 25, 41, 42, 43, 135, 139, 169, *185*, *187*, *188*, *194*

F

Ferwerda, F. P., 116, *188*
Frank, M. B., 1, 10, *191*

G

Gadeau de Kerville, H., 162, *188*
Ginsburg, B., 1, 92, *191*
Goldschmidt, R., 129, 152, 153, 181, *188*

S

Schlager, G., 25, 167, *191*

Schuurman, J. J., 116, 117, 136, 154, *191*

Scott, E. L., 165, *190*

Scott, J. S., 114, 161, *185*

Shaw, D. D., 13, 21, 89, 112, 113, 114, 121, 140, 161, *187, 192*

Shideler, D. M., 20, 25, 41, 43, 52, *186, 188, 189, 192*

Shrode, R. R., 2, 13, 15, 17, 21, 49, 122, 123, 127, 136, 154, 171, *194*

Shull, A. F., 111, 114, 115, *192*

Slater, A. J., 63, *192*

Slater, J. V., 63, *185, 192*

Slatis, H. M., 24, 91, 93, *194, 195*

Smith, S. G., 7, 8, 9, 157, *192*

Snow, R., 121, *192*

Sokal, R. R., 25, 168, 169, 172, *192*

Sokoloff, A., 2, 3, 13, 15, 16, 17, 19, 20, 21, 23, 24, 25, 30, 31, 35, 36, 37, 39, 40, 41, 42, 43, 45, 47, 48, 49, 51, 52, 53, 54, 55, 56, 57, 60, 61, 63, 66, 67, 68, 69, 70, 71, 72, 73, 76, 77, 78, 81, 84, 85, 86, 87, 89, 90, 91, 93, 95, 96, 97, 99, 100, 101, 102, 103, 105, 106, 107, 109, 110, 112, 121, 122, 123, 124, 126, 134, 135, 136, 137, 139, 142, 150, 151, 154 158, 162, 171, 172, 173, *186, 187, 188, 189, 193, 194*

Sonleitner, F. J., 106, 172, *192, 194*

Spencer, C., 100, *190*

Stanley, J., 1, 3, 24, 91, 92, 93, *194, 195*

Stanley, M. S. M., 96, *195*

Stay, B., 71, *191*

Sylvester, N. K., 113, 161, *191*

T

Tan, C. C., 111, 115, *195*

Tanaka, Y., 136, *195*

Timberlake, P. H., 115, *195*

Timofeeff-Ressovsky, H., 111, 115, 161, *195*

Timofeeff-Ressovsky, N. W, 115, *195*

Tobias, C. A., 63, *185, 192*

Todd, A. R., 106, *188*

Tower, W. L., 128, 129, *195*

V

Van Wyk, J. H., 106, *195*

Villee, C. A., 151, 152, *195*

von Heyden, L., 162, *195*

von Lengerken, H., 162, *195*

W

Waddington, C. H., 21, 30, 153, *195,*

Wagner, M., 100, *190*

Whittinghill, M., 136, *195*

Wolsky, A., 123, *195*

Woollcott, N., 106, *191*

Y

Yamada, Y., 20, *195*

Z

Zamora, R., 123, *195*

Subject Index*

A

Abbreviated appendage mutant of CS, 47

Abdominal mutants,
of CF, 84–85
of CS, 53–55

Abdominal structures, genes affecting, in CF, 96–97

Adalia, color variation in, 115

Ahasverus, body color mutant of, 112

Akimbo mutant of CS, 67

Alate prothorax mutant of CS, 50

Antennae,
genes affecting (*see* Antennal mutants)
and tarsi-fused mutant of CS, 77
and tarsi-reduced mutant of *T. molitor,* 118

Antennal mutant(s),
of *Cryptolestes turcicus,* 112
of CF, 85, 97–99
of CS, 33–34, 36–37, 40–42, 55–61
of *Latheticus,* 126
of *T. madens,* 110

Antennal segments, fused, mutant,
of CF, 97
of CS, 33, 36, 43, 55, 56
of *Latheticus,* 126
of *T. madens,* 110

Antennameres, fusion of in *Tribolium* and *Latheticus* mutants, 176

Antennapedia mutant of CS, 42–43

Anthonomus grandis, mutants of, 132–133

Appendages, mutants of (*see under* individual names of appendages)

Apricot mutant of *Anthonomous,* 132

Apterous mutant of *Bruchus,* 131

Arthrodactylus elongatus, teratology of, 76

B

Ballooned mutant of CS, 67

Banjo mutant of CS, 67

Bar-eye mutant of CS, 30

Bashful mutant of *Anthonomus,* 132

Bead mutant of CS, 61

Beetles (*see* individual species)

Bent elytral tips mutant of CS, 68

Bent femur mutant of CF, 103

Bent tibia mutant(s),
of CF, 103
of CS, 69
of *T. destructor,* 110
of *T. madens,* 110

Benzoquinone derivatives, in stink-gland fluid, 71, 106

Black mutant,
of CF, 91–92
of CS, 24–25

Blade elytra mutant of CF, 102

Blistered elytra mutant,
of CF, 99
of CS, 39

Body color,
genes affecting,
in *Ahasverus,* 112
in *Bruchus,* 131
in CF, 91
in *Cryptolestes,* 111
in CS, 50
in *Tenebrio,* 116
interactions of genes controlling, 27–30

* CF and CS are used throughout to indicate *Tribolium confusum* and *Tribolium castaneum,* respectively.

antennal, 174
of genitalia, 71, 103
Multiple allelic series,
of Bruchidae, 130–131
of Coccinellidae, 114–116
of *L. oryzae,* red, 121
of *T. castaneum,*
abbreviated appendages, 141
antennapedia, 141
black, 140
deformed legs, 141
fused antennal segments-3, 141
juvenile urogomphi, 140
pearl, 140
red, 140
Short antenna, 139
of *T. confusum,*
eyespot, 141
thumbed, 141
Mutants,
in Bostrichidae,
of *Rhyzopertha dominica,* black, 114
in Bruchidae
of *Bruchus quadrimaculatus,* 130
apterous mutation, 131
body color mutation, 131
eye color mutations, 131
macula mutation, 131
sex-linked inheritance, 131
somatic mutations, 131
in Chrysomelidae,
of *Gastroidea dissimils,* black and
green body color, 128
of *Gonioctena variabilis,*
color variation, 127
sexual dimorphism, 127
of *Leptinotarsa,* 128–130
(*Includes L. decemlineata, L.
defectopunctata, L. diversa, L.
melanicum, L. melanothorax,
L. minuta, L. pallida, L. rubi-
cunda, L. rubrivittata, L. signaticollis, L undecimlineata*)
of *Lina* (*see Melasoma*)
of *Melasoma scripta,* color variation
of elytra and thorax, 128
of *Phytodecta variabilis,* sex-linked
color pattern, 128

in Coccinellidae,
of *Adalia bipunctata,* affecting
elytral color, 115
of *A. decempunctata,* affecting
elytral color, 115
of *Cheilomenes sexmaculata,* affecting spotting and color pattern,
115
of *Coelophora inaequalis,* affecting
spotting and color pattern, 115
of *Epilachna chrysomelina,* "Divergens"-dominant with recessive
lethal effects, 115
of *Harmonia aulica,* spotting, 115
of *H. axyridis,* spotting, 115
of *H. conspicua,* spotting, 115
of *H. frigida,* spotting, 115
of *H. 19-signata,* spotting, 115
of *H. succinea,* spotting, 115
of *H. spectabilis,* spotting, 115
of *Hippodamia convergens,* elytral
spotting modifiers, 114
of *H. quinquesignata,* elytral spotting modifiers, 114
of *H. sinuata,* elytral spotting, 114
in Cucujidae,
of *Cryptolestes pusillus,* black form,
111
of *C. turcicus,*
crooked antennae, 112
red, 112
runty, 112
in Curculionidae,
of *Anthonomus grandis,* 132–133
(*Includes* apricot, bashful, ebony,
gnarled, milky, pearl, slate, yellow types)
in Dermestidae,
of *Dermestes maculatus,* 113
(*Includes* fuscous, light-wing,
pearl pleiotropic effects, rufous,
second sex pit, short elytra,
white types)
of *Trogoderma granarium,*
appearance in pupa, 113
effect on larval ocelli, 113
pearl, 113

Date Due

Demco 293-5